APPLIED LASER TOOLING

Applied Laser Tooling

edited by

O.D.D. SOARES
University of Porto, Porto, Portugal

M. PEREZ-AMOR
University of Santiago, Vigo, Spain

1987 **MARTINUS NIJHOFF PUBLISHERS**
a member of the KLUWER ACADEMIC PUBLISHERS GROUP
DORDRECHT / BOSTON / LANCASTER

Distributors

for the United States and Canada: Kluwer Academic Publishers, P.O. Box 358, Accord Station, Hingham, MA 02018-0358, USA
for the UK and Ireland: Kluwer Academic Publishers, MTP Press Limited, Falcon House, Queen Square, Lancaster LA1 1RN, UK
for all other countries: Kluwer Academic Publishers Group, Distribution Center, P.O. Box 322, 3300 AH Dordrecht, The Netherlands

Library of Congress Cataloging in Publication Data

Library of Congress Cataloging-in-Publication Data

Applied laser tooling.

  Includes index.
  1. Lasers--Industrial applications. I. Soares,
O. D. D. (Olivério D. D.) II. Perez-Amor, M. (Mariano)
TA1675.A67 1987   621.36'6   87-4135

ISBN-13: 978-94-010-8096-5   e-ISBN-13: 978-94-009-3569-3
DOI: 10.1007/978-94-009-3569-3

Copyright

Preface

The invention of the Laser, 25 years ago, has become an innovation with established industrial technology extended through diverse areas of economic viability (a 25% sales annual growth), and promising market perspectives.

In organizing an European Intensive Course on Applied Laser Tooling, it seemed opportune to bring together an international group of scientists to provide an appraisal of industrial Lasers, system integration, and sensitive areas of Laser beam material interaction, while emphasizing those areas which promise to have major impact both in science and technology.

Tutorial papers and reports on latest developments both in research and industrial manufacturing were complemented by video and film projections to show the wide variety of applications in industry, stressing the combination of Lasers with other technologies, mainly CNC and Robots.

The large participation by the industry fulfilled the intended interaction and cross-fertilization between the scientific, technological and industrial community, reinforcing the innovative capacity readily demonstrated at panel discussions.

It was neither possible nor planned to cover all the aspects in full depth. Efforts were addressed to selected areas where discussion of advanced knowledge and technology topics would stimulate further progress of Laser tooling (in main directions: software, hardware and peopleware).

Laser tooling was then discussed in light of its major applications covering Laser beam robotic manipulation towards flexible manufacturing systems.

The following articles give a fair account of the course programme.

Some of the texts were written from lecture notes. While the writers asked the lecturers to review the material, the Editors will bear an unexpected enlarged responsibility.

At closure there was a sense of excitement towards future possibilities mixed with pleasant recollections of the friendly atmosphere enjoyed at Samil.

This volume attests the pleasure of sharing this experience with others interested in the field.

Olivério D.D. Soares
Mariano Perez-Amor

Samil, Vigo, 1986

Sponsorship and Acknowledgements.

The Editors and Directors of the Applied Laser Tooling Course wish to thank the Council of Europe, the Director of the European Intensive Course programme, as well as the other institutions that equally sponsored the course in various forms such as providing the required financial resources so indispensible for its preparation and organisation.

- Higher Education and Research Division - Council of Europe
- E.P.A. - European Photonics Association
- Dirección Gral. de Innovación Tecnológia - Ministerio de Industria
- Conselleria de Industria, Comercio y Medio Ambiente - Xunta de Galicia
- CAICYT - Comisión Asesora de Investigación Cientifica y Técnica
- Ministerio de Educación y Ciencia
- Universidad de Santiago
- Excom. Ayuntamiento de Vigo
- The British Council
- Divisao de Optica da Sociedade Portuguesa de Fisica
- IBERIA - Spanish Airlines.

CONTENTS

VIII

APPLIED LASER TOOLING

Olivério D.D. Soares
Applied Physics, Faculdade de Ciências, Universidade do Porto,
4000 Porto, Portugal

Mariano Perez-Amor
Applied Physics, E. T. S. Industriales, Universidade de Santiago
Apartado 62, Vigo, Spain

Abstract

Laser technologies are credited as one of the essential components of present pattern of the industrial competitive transmutation. Companies to respond with flexibility and efficiency are led to the technological combination of Laser, Robots and Optical Fiber systems to increase competitiveness through high productivity, innovation and quality assurance. Industrial and economical relevant aspects of the Laser era are dealt with while emphasizing the Laser processing of metallic and non-metallic materials, microelectronics fabrication and processing of semiconductors.

Soares, O.D.D., Perez-Amor, M. (eds), Applied Laser Tooling. ISBN-13: 978-94-010-8096-5
© *1987. Martinus Nijhoff Publishers, Dordrecht.*

1. Lasers and Industrial Dynamic Competition

"By the XXI century the photon will replace the electron as the element of technological development"

Guy Denielon
ANTRT - Association Nationale de la Recherche Technique

Power Lasers have already been around for a considerable time but only recently came the expanding use of their full potential for manufacturing. Many manufacturers seem rather cautious of introducing Laser technology into their factories which may be due to yet a certain mystery around the topic. The present book hopefully will give some contribution to dispel these hesitations and bring Laser technology to the understanding of their benefits and limitations to reluctant industrialists.

The study and development of Laser applications, namely in material processing is in constant development further increasing the perspectives to the industrialist. The Laser relevance derives from its innovative capacity to produce:

    i) high technology
   ii) high knowledge
  iii) flexible energy and information handling

with consequent adaptability to manage the change in the process of transfer of science into economically productive technology, based on the recognized essential trends:

    i) high productivity through high production rate and complete quality
       control cycle.
   ii) innovation to enforce the dynamic comparative advantages, essential
       ingredients to the economic recovery and growth
  iii) miniaturization for flexibility, adequacy, and saving of raw
       material and energy.

These are the progress dynamic vector components, to overcome the present crisis, and promote a sound and stabilized development of the new economy progressively technology oriented.

These progress components proceed from the need for an economical improvement through the evolving market competition that is usually linked to productivity features (quality and rate), and innovation.

However, innovation is a matter of great controversy, particularly when looking at high technology (a hopeful portent for eventual economic recovery and growth). An aspect then discussed further on is economics of Laser applications. This has been brought into the course intentionally, to make it to emerge the difference between invention, discovery, and innovation (1), notably when technology and science are becoming a service beyond a mere product. Further, it should be emphasized that innovation far from being peripheral must be considered a powerful though complex driving force that works its way through the economy and molds it, Fig. 1. Its success is anchored on a dynamic competition rather than a consequence of static economic growth with a pure reproduction of products or services. It

GROWTH OF AN ENTERPRISE AND FINANCING SOURCES

	INCUBATION PERIOD		IV LAUNCH	V INDUSTRIALIZATION	VI MATURITY	VII DECLINE OR
I) CONCEPTION (Dining room table)	II) PROJECT	III) DEVELOPMENT (Laboratory/Garage)	(Start - up)	(Lift - off)		REJUVENATION

Fig. 1: Economical cycle of innovation (12)

Fig. 2: Stroboscopic and Phase-modulated Moiré-holography set-up as an Optical Metrology technique for vibration studies (Applied Physics Laboratory - University of Porto)

is the competition within industries engaged in using renewed technological means that governs how rapidly new technologies are incentivated to evolve (another present case is the microprocessor market). This dynamic comparative competition increases, in principle, both the wealth of nations and their economic stability.

Dynamic competition demands the handling of two kind of risks:

i) technological risk that derives from attempts to overcome limits of some kind to innovate.
ii) the danger of being overtaken by a competitor

These are components of the actual profile of the transmutation of science into economically productive technology. Furthermore, this stiff worldwide competition is coupled to a change of the:

i) open world technology (man-robot-computer-laser-machine)
ii) economic and political pattern

and is associated with an international redistribution of production and investment.

As a pertinent example, Robots and Laser developed independently in the past are now combined in a wide range of systems providing spatial flexibility and accuracy, associated with process flexibility - a tool that cuts, welds, heattreats and drills a broad range of parts and materials.

Laser tooling forms a part of the flexible manufacturing and non-manufacturing concept.

2. Laser Technology

On its 25th anniversary (Table I) the Lasers today hold a secure market position derived from their capabilities (Table II) to play a part in all stages of manufacture, from the raw materials processing to finish operations assembling, inspection and product quality control.

The unique characteristics of Laser radiation (3):

i) Coherence (spatial, temporal)
ii) Divergence
iii) Energy (power, focusing)
iv) Mode structure
v) Polarization
vi) Pulse rate or CW
vii) Wavelength

determine the increasing number of applications of Laser.

Artificially one may consider two main aspects of interaction of radiation: information and energy. The information domain relates to processes and corresponding techniques for metrology, sensing, processing and control (4), (Table III), serving a vast domain of innumerable applications, Fig. 2. These applications result from the fact that Laser radiation can, in principle, be entirely controlled in amplitude, phase, monochromaticity, coherence, polarization, directivity, the Laser beam

TABLE I

LASER HISTORIC (6)

LASER TYPE	PUBLICATION	LABORATORY
IR	A.L. Schawlow and C.H. Townes Infrared and Optical Masers Phys Rev. 6 (1958) 112	Bell Columbia
SOLID	T.H. Maiman Stimulated Optical Radiation in Ruby Nature, August 6 (1960) (Report results from 16th May 1960)	Hughes
GAS	A. Javan, W.R. Bennet, D. Herriot Continuous Maser Oscillation in a Gas Discharge with He-Ne Mixture Phys. Rev. Lett. 6 (Feb 1961) 106 (Rec. 10th Dez. 1960)	Bell
SEMICONDUCTOR	R.N. Hall, G.E. Fenner, J.D. Kingsley, T.J. Soltys and R.O. Carlson Coherent Light Emission from Ga As Junctions Phys. Rev. Lett. 9 (1962) 366 (Rec 24th Sept 1962)	G.E.C.
DYES	P.P. Sorokin and J. Laukard Stimulated Emission in Dyes IBM Journal R&D 10 (1966), 162	IBM

TABLE II (2)

LASER USAGE SAMPLING

RESEARCH	SOLID STATE (Glass-Crystal)	$Al_2 O_3 Cr^{3+}$; YAG Nd; Glass: Nd; YAP: Nd; YAG: Er G SGG, Cr Nd, Alexandrite
	GAS	Noble Gas: He-Ne; Ar^+; Kr^2; Xe^+ Metallic Vapour: He-Cd; He-Se; Cu; An Mollecules: CO_2; CO; HF; DF Excimer: $X_e F$; Kr f
	LIQUID	Dye: Rhodamine GG; B; Coumatine
	SEMICONDUCTOR	Ga As; Ga Al As; Ga As P Pb, Sn, Te
	PICOSECOND LASER	Ring lasers
	INDIRECT LASERS	Laser pumped media Gaseous, Raman Shifted NL SHG, THG, OPO
	NEW SOURCES	X-Ray Laser Free Electron Lasers
SCIENTIFIC INSTRUMENTATION	NON-LINEAR OPTICS	Harmonic Generation Optical Parametric Oscillations Raman Effect (spontaneous and stimulated) Brillouin Effect (spontaneous and stimulated) Four-wave-mixing Optical Phase Wave Conjugation
	SPECTROSCOPY	High Resolution Spectroscopy Non-linear Spectroscopy Saturation Spectroscopy
	PHOTO-INTERACTION	Thermonuclear Fusion Selective Photochemistry Isotope Separation
	BISTABILITY	Logic Circuitry Processors Computer Sensors
	SYNERGETICS	Transition Order-chaos
INDUSTRIAL TOOL	METROLOGY	Length Standards Distance Measurement Telemetry and Triangulation Anemometry Dopler Remote Monitoring Optical Radar Alignment Sensors Survey: Inspection, Quality Assurance and Control

		HOLOGRAPHY Display - transmission or reflection
		    white light rainbow hologram
		    Photoresist - Reflection
		    Embossed,Head-up displays
		  Optical Elements - scanners, deflectors, viewfinders
		  Non-destructive Testing
		    - Interferometric Holography
		    - Hybrid Fringe Analysis
		  Crystal Growth

		SPECKLE Speckle Monitoring
		  Speckle Photography
		  E.S.P.I.

	MATERIAL PROCESSING	Metal Working: cutting, welding, drilling, hardening, cladding, alloying, marking Non-metallic Material Processing: wood, plastic, textile, glass, leather, wood Microelectronics Semiconductors Annealing Photochemistry LCVD Selective Processing - photodissociation, photoionization
	ENVIRONMENT PROTECTION	Analysis Pollution: LIDAR Oceanology
	MEDICAL	Surgery Photochemiotherapy
	BIOTECHNOLOGY	Selective Photochemistry Picosecond Events
	TELECOMMUNICATIONS	Optical Aerial Links Optical Fibres - network, submarine cables, LAN Integrated Optics
	INFORMATICS	Data-handling Processors Memories Computer Printing and Graphics Art
	MILITARY	Range Finder Target Designator Tactical Weapons

TABLE III

LASERS - INFORMATION

INFORMATION	Acquisition	Measurement Transducers Sensors Non-destructive Testing Inspection, Quality Assurance and Control Survey
	Storage	Data-handling Memory
	Transmission	Communications - Optical Fibres, Integrated Optics Telemetry Printing and Graphics
	Processing	Image Processing Logic Circuitry Computing Art
	Presentation	Display Scanners and Deflectors Viewfinders Holography

taking the place of carrier or reference with obvious key advantages:

   i) Laser beams do not interact
  ii) Optical channels are well adequate for parallel processing
 iii) Optical channels are capable of very-high speeds (femtoseconds!)

The Laser radiation can efficiently and reliably be integrally controlled in intensity and focusing. This controllability, accuracy and reliability of Laser sources are the core of Laser applications in material processing, and is considered under this title of Laser Tooling, Fig.3.

Each application requires its own parameters adjustment in regard to power and energy levels, per wavelength, beam profile and modulation method. The most important being the balance between power and good beam quality, resulting in small focused spot size and highest energy density. Then, optimized use can be done of the advantages of Laser Tooling:

   i) beam parameters control with consequent tight control of manufacturing process
  ii) negligible forces on the pieces
 iii) easy of adaptation to automation
  iv) sharing of movement between Laser delivering head and piece to the optimization of production

Traditional methods of material processing are dependent on delivery of various forms of mechanical, electrical or chemical energy into contact with the workpiece. The availability of direct energy sources (5): Lasers, electron beam, and ultrasonic devices, capable of power densities in excess of 10^6 W cm^{-2} , introduced the non-contact tooling concept. Knives, drills, abrasive wheels, flames, chemicals and electrodes seen in certain industrial operations, can be eliminated, while reducing maintenance, replacement and direct labor costs.

The behavior of materials subjected to intense concentration of essential thermal energy up to 10^5 W cm^{-2} remains of central interest, in particular, those correlated with machining, welding, and surface engineering (leaving e.g. refractive index changes, electrical conductivity changes, etc).

Laser performances continue then to be explored, to established a full but still developing generation of purpose-built family of Lasers for industry, Table IV.

In parallel, concentrated efforts have been focused on system development of better automated techniques and cost-effective solutions competing with existing methods in industry.

As a result Laser technology is also growing as a horizontal technology i.e. contributing to research, development and production, percolating into almost every domain. There is then no wonder that Laser sales present an annual 35% sales growth rate.

3. Laser Tooling

Laser tooling has been used to define the answers of Laser Technologies to the production systems.

Market acceptance of industrial Lasers for production floor machining

TABLE IV

TYPICAL LASER CHARACTERISTICS

TYPE	WAVELENGTH [μm]	MODE OP.	MAX REPT. RATE [Hz]	PULSE WIDTH [ms]	MAX FOCUSED ENERGY [J cm⁻²]	MAX FOCUSED POWER [W cm⁻²]	MODE STRUCTURE	POWER TEM00 [W]	AVERAGE POWER [W]	PULSE ENERGY [J]	TYPICAL USES
He Ne	0.6328	CW					TEM00	$5-50 \times 10^{-3}$			Metrology
Ar	0.4519 to 0.5145	CW					TEM00	2-40			Holography Raman Spectroscopy
Ar	0.4880 or 0.5145	CW					TEM00	0.5-10			Semicondutor Annealing
CO_2	10.6	CW					MM	500	Up to 10^5		Mat. Processing
CO_2	10.6	CW				10^7	TEM00		10^3		"
CO_2	10.6	CW					MM		2×10^4		"
CO_2	10.6	PULSED (DP)	$< 2.5 \times 10^3$	> 0.1	10^4	10^8			300	< 1	"
CO_2	10.6	PULSED (QS)	$< 3 \times 10^5$	$10^{-5}-3 \times 10^{-4}$	10^3	10^{10}	TEM00		100	< 1	"
CO_2 TEA	10.6	PULSED	400	0.4					10^5		"
He Cd	0.325	CW									Photo lithography
KRYPTON	0.3597 to 0.7993	CW						6			Holography Raman Spectroscopy
NITROGEN	0.3371	PULSED	to 500	10^{-3}			MM		2.5×10^5	2.5×10^{-3}	Photochemistry
Nd-Glass	1.06	PULSED	1	0.5-10			MM		10^6	0.125	Mat. Processing
Nd-YAG	1.064	CW				2×10^7	TEM00	20			"
Nd-YAG	1.064	CW				4×10^6	MM	400			"
Nd-YAG	1.064	PULSED FLASH LAMP	< 400	0.1-10	5×10^5	10^9	MM		400	< 50	"
Nd-YAG	1.064	PULSED (QS)		$2 \times 10^{-2}-0.3$	10^5	3×10^{11}	MM		150	< 10	"
Nd-YAG	1.064	PULSED (QS)	5×10^4	0.2			MM		50	5×10^{-3}	"
Nd-YAG	1.064	PULSED (DP)	< 400	0.1-10	5×10^6	2×10^{10}	TEM00		20	< 5	"
RUBY	0.6943	PULSED								1-500	Welding Drilling
RUBY	0.6943	PULSED (QS)							$1-10^{10}$		Pulsed Holography Raman - Brillouin, Vaporization

Fig 3: A CO_2 Laser (CW - pulsed) work station (E.T.S. Industriales, Univ. Santiago, Vigo)

has shown a 35% growth (2) (30,000 units by the year 2,000) with materials processing the fastest growing segment.

Machine-tools suffer from wearing, slow rate of production and the need for a large stock of tools. Lasers can emit an intensive, highly collimated beam that can be adjusted to perform most of the operations on material processing:

    i) Welding (melting)
    ii) Material Removal (drilling, trimming, evaporation, sputtering)
   iii) Material Shaping (cutting, scribing, controlled fracturing, marking)
    iv) Thermally Induce Change (localized heattreating annealing, and surface oxidation, grain size control, diffusion, zone melting, cladding, alloying)
    v) Chemical Induce Change (photochemical reactions)

to name the most important.

Lasers are highly adaptable to automation and micro-machining control. Laser (high speed, precision, cleanness) and robot technologies (high accuracy, speed, and large spatial envelope) have now been combined in a major step in the advancement of flexible manufacturing system. The Laser/robot system has the spatial flexibility and accuracy of a robot and the process flexibility of a Laser. A single Laser/Robot system can cut, weld, heattreat and drill a broad range of parts and materials.

Laser tooling then is progressively taking a prominent position as a manufacturing tool for its: high-precision, life-time and specific characteristics (8):

- clean, optical energy
- availability of high density energy (10^6 W/cm^2)
- focused localized, small size (several wavelengths) interaction area with comparative negligible heating and distortion of surrounding material
- stable and quiet interaction with workpiece leading to minor after machining
- non-contact tool (no contamination, no wear, no corrosion, or breaking, no mechanical forces, exact positioning, precision, simpler fixturing)
- Tooling, fixtures, setup, and inspection, in general, relatively simple and easily adaptable to production procedures
- multiple operation in one programme cycle without retooling
- exact and flexible controllability of the processing, dimensions and quality; high reproductibility
- processing in difficult to reach areas, simplification of workpiece design and handling by beam deflection and beam splitting
- numerical control permits design of unique geometries, to shape materials that mechanical operations cannot
- processing in any atmosphere, including vacuum
- ideal possibilities of robotization and automation
- maximum degree of exploitation derived from high reliability of Laser system and the availability of on-line process control
- Laser data to achieve a desired processing in a given material can be predetermined
- process complex shapes at high speed (combined with computer controlled positioning systems)
- burn-free edge shapes (while simultaneously sealing edges for

composite materials), eliminating or significantly reducing post-processing steps
- elimination of the cost and delay of special tooling, critical for small quantities of special shape cuts
- high accurate metal cutting with narrow kerf width and minimal heat affected zone
- environmental clean (less noise, dust, swarf and lubrificants)

Laser tooling, in consequence, shows an impressive potential ability to material processing in rather innovative way resulting in a radical change of exigencies and satisfaction both of production and employment.

In many material processing applications, Laser techniques are more precise, cheaper, and faster than other alternatives. In some cases they are even the unique solution (advanced composite, plastics, ceramics and exotic alloys) and, in general, they present the proper compatibility with control technology of modern digital electronics (9).

The majority of present installations are stand alone equipment or machine integrated into an assembly line, usually dedicated to one product or process based on economic justification by a request for high-volume, long-run production (to offset relatively large initial cost). The trend is however towards flexible manufacturing - short runs and product variety. The Laser is then computer controlled and performs a multitude of operations. The versatility of Laser technology is then adjusted to the industrial application in terms of power/energy level, wavelength, beam profile, and modulation technique. The major step in integration of Laser, indeed the forefront of manufacturing technology was achieved with its combination with robots, Fig. 4.

Integrated system, Laser/robot are being introduced in flexible manufacturing systems. This is further developing Laser technology both in beam delivering systems and light weight and compactness.

Laser tooling is then a new solution for the transmutation of manufacturing technologies and for novel tooling techniques introduced to respond to the requirements of productivity, reliability and quality of new products or new designs.

Once a Laser solution has been considered research is conducted to:

- establish a mathematical model how to determine in first approximation the Laser characteristics to meet a predetermined processing at a given workpiece material
- find suitable Laser system fulfilling these requirements
- study the process control to judge the economic technical feasibility of the use of Laser for the given problem as compared with other processing techniques.

4. Material Processing Operations

Laser radiation is centered at a characteristic wavelength and provides an intense collimated beam that can be focused to a diffraction limited spot.

Mode structure defines the beam divergence and intensity profile. The fundamental TEMoo mode is of Gaussian distribution, minimum divergence, and

12

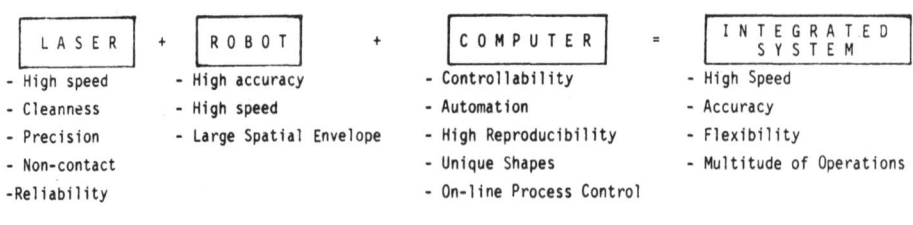

Fig. 4: Laser System Metamorphosis

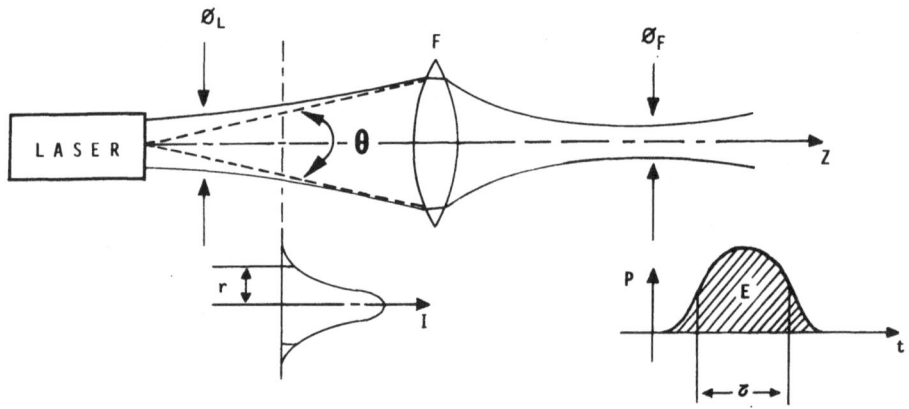

Fig. 5: Laser beam, TEM$_{00}$ mode.

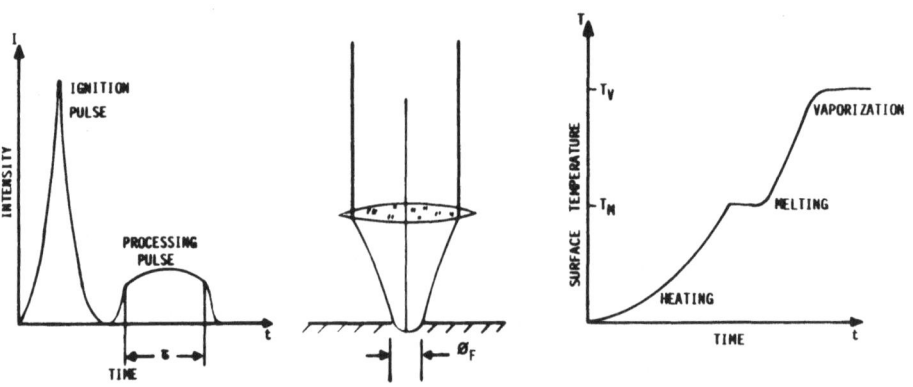

Fig. 6: Surface laser heating response (8)

maximum energy flux, Fig. 5:

$$\theta_{min} = 1,22 \ \frac{\lambda}{R} \tag{1}$$

$$P = \frac{4 \ E}{\pi \ F^2 \ \theta^2 \ \tau} \tag{2}$$

where:
P = power density at focal plane of the lens
E = energy output from the Laser
F = focal length of the lens
θ = beam divergence (full angle)
τ = laser pulse length

While there is a broad range in specific characteristics (e.g. wavelength) of commercially available Laser, CO_2 - Lasers and Nd-YAG-Lasers are the mainly used. Applications requiring high average power use CO_2 - Lasers and for high energy per pulse Nd-YAG-Lasers are preferred.
In a Laser processing system five subsystems can be identified:

i) The Laser consisting of:
- Laser head (active medium, excitation source, power supply, cooling unit)
- Resonant cavity (mirrors, aperture and other mode structure confining elements)
- Modulator (beam modulation or pulse shaping)
- Beam monitoring (power meter, mode evaluation, pulse duration, feedback loop)
- Optical isolator (stop of retroreflected radiation)
- Polarizing Component
ii) Beam guiding and shaping (optical system, optical waveguides - optical fibers, deflection system - $\simeq 1\ 000$ cm s^{-1}, moving arm - $\simeq 50$ cm s^{-1}, moving piece $\simeq 10$ cm s^{-1})
iii) Workpiece positioning and handling (coordinate table, robot)
iv) Integration and control unit (computer, robot, artificial intelligent system)
v) Safety enclosure (rugged enclosures for optical, mechanical and electronic system, protective enclosure of work station)

What happens when a Laser radiates a material depends on the characteristics of the Laser beam, of properties of the material (optical, thermal, surface state) and optimization of transfer of energy (absorption capacity and rate, heat conduction). In broad terms, Fig 6 and 7, a few guiding rules can be used as first approach (10):

i) A high intense initial pulse (short width) overcomes the effect of high reflectivity of smooth metallic surfaces

POWER DENSITY $\left[\text{W cm}^{-2} \right]$	PULSE LENGTH $[\text{s}]$	TEMPERATURE REGIME	PROCESSING OPERATION
10^5	0.1	$T < T_M$	Transformation Hardening
10^6	2×10^{-4}		Semiconductor Annealing
10^7	1.5×10^{-8}		
10^6	$10^{-3} - 10^{-2}$	$T_M < T < T_V$	Surface Melting - Alloying - Soldering Welding by Axial Heat Cond.
10^7	$10^{-3} - 10^{-2}$		Deep Penetration Welding
10^8	$0.1 - 0.5$	$T > T_V$	Drilling, Cutting Scribing Trimming

Fig. 7: Processing operations vs temperature regimes from combined intensity and pulse length (8)

ii) Power density and pulse duration are adjusted to obtain the required process: heating without melting; melting or vaporization

iii) Lateral dimension of processing area is mainly dependent on beam spot size (μm range)

iv) Depth of interaction depends on heat conductivity, and material removal, being controlled by beam pulse intensity and width.

The incident energy is absorbed according to Lambert's law:

$$I(z) = I(z=0)\ \exp(-\alpha z)$$

where:

$I(z)$ = radiation intensity after a propagation depth z.

α = absorption coefficient

Radiation absorption is mainly a surface effect in the materials of major interest within a depth ranging from 0.1 to 1 μm. The heat then propagates into the bulk material essentially by heat conduction (5).

For an isotropic solid, the heat flow $\vec{f}$, $\left[J,\ s^{-1}\ cm^{-2} \right]$ is given by (11):

$$\vec{f} = -K\ \text{grad}\ T$$

where:

K = coefficient of thermal conductivity $\left[W\ cm^{-1}\ K^{-1} \right]$

T = absolute temperature $\left[°K \right]$

The energy conservation continuity equation is then:

$$\text{div}\ \vec{f} + \rho c\ \frac{\partial T}{\partial t} = 0$$

where:

ρ = mass density $\left[g\ cm^{-3} \right]$

c = specific heat $\left[J\ g^{-1}\ K^{-1} \right]$

t = time $\left[s \right]$

The diffusion equation takes the form:

$$\frac{\partial T}{\partial t} = k\ \Delta T$$

where:

$k = K/\rho c$ is the thermal diffusivity $\left[cm^{-2}\ s^{-1} \right]$

That can be written in terms of heat flux:

$$\frac{\partial \vec{f}}{\partial t} = k\ \Delta \vec{f}$$

In the presence of a heat source characterized by a production rate S $[J\ s^{-1}\ cm^{-3}]$ it comes:

$$\rho c\ \frac{\partial T}{\partial t}\ =\ K\ \Delta T + S$$

In the steady state:

$$\Delta T = \begin{cases} -S/K \\ \\ 0 \end{cases}$$

Solutions of the diffusion equation for one-dimensional flow are given in ref 11 by Carslaw and Jaeger. For a Laser heating of a circular area of radius ϕ on the surface of a semi-infinite medium, at a point z on the axis perpendicular to the heated circle and passing through its center:

$$T\ (z,t) = (2\ F_o\ \sqrt{kt}\ /\ K)\ \left\{ ierfc\ (Z/2\ \sqrt{kt}) - ierfc\ \left[(Z^2 + \phi^2)^{\frac{1}{2}}\ /2\ \sqrt{kt} \right] \right\}$$

where:

F_o is the incident flux $[W\ cm^{-2}]$

$ierfc\ z = \int_z^\infty erfc\ U\ du$

$erfc\ z\ =\ 1-erfz$

$$erf \equiv 2/\sqrt{\pi}\ \int_o^Z e^{-\xi^2}\ d\xi$$

These equations require corrections for the energy carried away by emerging vapors.

Several authors have drawn approximations as results for diverse situations.

Under Melting Processing (8)

i) Hardening

Intensity and pulse width control of the Laser beam allows the heat treating of thin layers of the workpiece surface. High cooling rates results from self-quenching by heat conduction. Solid phase growth velocity, vg, depends on temperature and crystal orientation $(0.1\ cm\ s^{-1}$ near melting temperature). Transformation hardening (hard and wear resistant surface structure) varies with carbon content, crystal orientation and metallurgical structure. Laser parameters for a surface layer of z [cm] depth heating to a temperature Th [K] and a spot size ϕ[cm] are given by:

	MAIN PROCESS	
	Heat Conduction	Solid-phase growth velocity
Power Density	$Ia = \dfrac{K}{Z}(T_M - T_h)$	$Ia = \dfrac{T_M\,K}{Z}\sqrt{\dfrac{\pi}{k}\dfrac{Vg}{Z}}$ $[\text{W cm}^{-2}]$
Pulse Length	$\tau = \dfrac{Z^2\,\pi}{4k}\dfrac{T_M}{T_M - T_h}$	$\tau = \dfrac{Z}{Vg}$ $[\text{s}]$
Spot Size	$\phi_F \simeq \phi$	$\phi_F = \phi$ $[\text{cm}]$

where:

  Vg = solid - phase growth velocity $[\text{cm s}^{-1}]$ at temperature T_h.

At depth z the cooling rate is approximately:

$$-\frac{\partial T}{\partial t} = \frac{k}{z\sqrt{\pi}} = \frac{(T_M - T_h)^2}{T_M}\,\frac{T_h\,Vg}{2}\left(1 - \frac{Vg\,Z}{2k}\right)$$

Laser power P and scanning speed V are obtained from:

	CW Radiation	Repetitive Pulse Radiation
Power	$P = Ia\,\dfrac{\pi}{4}\,\phi_F^2$	$P = Ia\,\tau\,\dfrac{\pi}{4}\phi_F^2\,\nu$ $[\text{W}]$
Speed	$V = \dfrac{\phi_F}{\tau}$	$V = (1-S)\phi_F\,\nu$

where:

  ν = repetition rate $[\text{Hz}]$
  S = degree of spot overlap

ii) Annealing (semiconductors)

Annealing of crystals defects from ion-implantation in semiconductors is done using Laser processing below melting point to cope with requirements for submicron resolution.
Continuous Argon Lasers are generally used at an energy density of the order of 200 J cm^{-2}.

iii) Controlled Fracturing (10)

Controlled fracture process is a technique for separating materials (e.g. brittle materials). The absorption of the Laser radiation induces thermal gradients producing mechanical stresses sufficient to fracture the material over such a small region that it does not propagate uncontrollably.

Above Melting Processing (8):

i) Surface Melting, Alloying

The melting of thin surface layers requires a Laser beam with:

Power density: $Ia = \dfrac{K}{2} (T_V - T_M)$

Power Length $\tau \simeq \dfrac{\pi}{4} \dfrac{Z}{k} \left(\dfrac{T_V}{T_V - T_M} \right)^2$

Spot size $\phi_F \simeq \phi$ $(\phi >> Z)$

For larger surfaces or specified contours the Laser beam is scanned by appropriate deflection system and modulators.
Surface melting leads to extremely hard, wear resistant surface with improved fatigue response.
The high surface tension of liquid metals results in smooth surfaces. Vapor deposited or sprayed addition material on the melting surface originates an alloy with additional properties.

ii) Microsoldering and microblazing

High intensity and short pulse duration of Laser beam combined with its compatibility to beam deflection and computer control integration turns Laser microsoldering and microbrazing a cost-effective tool for large production industry manufacturing.

iii) Semiconductor Annealing

Recrystalilization in liquid phase at very high speeds (Z m s^{-1}) is used in semiconductor damage annealing.
Grain size enlargement of doped poly-silicon layers deposited in

oxides is also performed with Laser surface melting.

iv) Welding or Joining

Laser welding offers higher quality, greater accuracy, higher
production rates than conventional methods and compatibility to
join dissimilar metals. In present it is the widest use of Laser
material processing.
At high power densities the material vaporizes in the center of
Laser beam. The widely called key-hole is formed as a balance
between power, vapor pressure, melting, and radial heat conduction.
Optimum quality melt pool of Z depth and ϕ diameter is formed for:

Power Density
$$I \simeq Z \frac{T_V K}{\phi^2} \left[1 + \left(\frac{T_V}{T_M} \right)^2 \right]^2$$

Pulse Length
$$\tau \simeq \frac{4 \phi^2}{k} \left[\frac{T_V^2 - T_M^2}{T_V^2 + T_M^2} \right]^2$$

Spot size
$$\phi_F \simeq \frac{2 \phi}{1 + \left(\dfrac{T_V}{T_M} \right)^2}$$

within the approximation of weld pool diameters within $Z \geqslant \phi \geqslant 2/8$.
For dissimilar metals metallurgical criteria from conventional
techniques can be applied. Welding additives are also applied in
the form of foil, wire and powder.
Non-metallic materials (plastics, glass compounds, etc) can benefit
from Laser welding and at lower power densities flexibility is
increased.

Above Vaporization Processing (8)

Sufficient increase of Laser power density causes material removal by
vaporization and melt ejection. Machining operations (drilling, cutting,
milling, scribing, trimming, etc) can be Laser performed in hard and
brittle material, metallic and non-metallic, in a microtechnique of
precision and accuracy as well as in large mechanics.
Edges quality and redeposition of ejected material establish the
finish of Laser machined pieces.

i) Drilling

High vaporization rate is requested to remove the material leading to power densities in the 10^7 - 10^8 W cm^{-2} range. Higher intensities may result in heat and pressure shock in the vapor cloud reducing drilling efficiency and quality. For drilling shallow holes $(4\ Z\ /\ \phi \leqq 1\ )$ with diameter ϕ [cm] and depth Z [cm], required Laser characteristics are:

Power Density $\quad I_o = 10^7\ -\ 10^8\quad W\ cm^{-2}$

Pulse length $\quad \tau\ =\ Z\ \dfrac{L_v}{I_o}\quad (1 + 0.35 \sqrt{\dfrac{k\ L_v}{Z\ I_o}}$

Spot size $\qquad \phi_F = \phi$

For the drilling of deep holes $(4\ Z\ /\ \phi \gg 1)$ it can be used:

Power density $\quad I_o\ =\ 10^7\ -\ 10^8\quad W\ cm^{-2}$

Pulse length $\quad \tau\ =\ Z\ .\ \dfrac{L_v}{I_o}\quad \dfrac{\phi\ -\ 0.2\ \sqrt{\dfrac{k\ L_v\ Z}{I_o}}}{\phi\ -\ 0.7\ \sqrt{\dfrac{k\ L_v\ z}{I_o}}}$

Spot size $\qquad \phi_F\ =\ \phi\ -\ 0.7\ \sqrt{kZ\ \dfrac{L_v}{I_o}}\ .$

For thermal sensitive materials a multitude of short spikes (≈ 0.3 μs) is used to remove layer by layer without disturbance in the surrounding material.

ii) Cutting, Scribing, Milling

Cutting width and lateral heating can be controlled through Laser beam characteristics either at CW and Pulsed operation, combining the relative movement of Laser beam and work piece.
The Laser power and advance speed for deep cutting $(4\ Z\ /\ \phi \gg 1)$ are approximately:

Fig 8: Laser Processing Economic / Technical Feasability

COST	Specific Application and Exploitation Degree		Favorable Pattern
Investment	- Power - Flexibility - Complexity of Handling System	- High Production Rates	
Maintainance	- Laser Active Medium - Laser Pumping System - Optical Components	- Tool Wearing Expensive - Scrapping and Residual Distortion to be Avoided	
Operation	- Consumables (gas, electricity, cooling, water) - Personnel	- Automatization for considerable Cost Reduction, or Impossible with Conventional Means - Processing Operations and Quality Improvement Required	

	CW RADIATION	PULSED RADIATION
Power	$P = 3.7\ Z\ L_v\ k$	$P = 0.5\ Z\ L_v\ \phi^2\ \nu$
Speed	$V = 7\ \dfrac{k}{\phi}$	$V = (1-S)\ 0.7\ \phi\ \nu$
Spot size	$\phi_F = 0.7\ \phi$	$\phi_F = 0.7\ \phi$

Maximum thickness to width rate ranges $Z/\phi = 10$ limited by melt expulsion.
Coaxial or inclined gas jets are applied to protect the focusing lens and to aid the material removal. The gas can be also used to activate or inhibit reactions with the material.

Laser Induced Chemical Reactions (10)

Focused Laser energy can be used to initiating certain chemical reactions (ionization, gas breakdown, etc)

Laser processing referred applications are far from exhausting the present uses of developed Laser material processing techniques. They show however that the past has been fairly rich, and the future promise a realist and rapidly expanding broad field of Laser manufacturing techniques in industry.

5. Conclusion

Lasers so far were mainly used as substitutes of conventional processing tools in order to improve production rate and quality, or to reduce production cost, and handle difficult materials or complex manufacturing operations. Continuous developments of Laser system and process by progressive knowledge of Laser material interaction are producing the enlargement of application domains and the establishment of novel cost-effective fabrication techniques.
The time has come for a rapid growth of Laser uses in manufacturing systems, after the development and demonstration of reliable industry-rated Laser sources, optical beam-delivery, control systems and safety procedures (13).
While the specific characteristics of Lasers make them the ideal tool for high precision mass production and conventionally non-soluble

manufacturing problems (8), economical/technical feasibility of Laser techniques are far reaching, Fig 8, and Laser material processing is in many case the most favorable solution.

Present advances of CNC makes it possible to create a laser machine center capable of drilling, cutting and spot welding the same workpiece in one programmed cycle (14).

The combination Laser - robot and optical fibre will further push the industrial development of Laser material processing.

Eventually, the intelligent Laser systems will make it possible the direct communication with Laser and computer-numerical-control devices. The Laser beam will then show programmed characteristics precisely defined for the specific process.

High technology markets such as Lasers, with a broad range of secure and potential applications are difficult to quantify. However, recent surveys of both EEC and world markets for Laser Systems predict very strong growth over the next decade, especially in the areas of data handling, communications, materials processing, medicine and semiconductors fabrication (13).

Acknowledgments

This article profited of interaction with colleagues at several meetings and is partially based on collected material, experience and stimulating discussions from the EIC, in Samil-Vigo.

References

1. O.D.D. Soares, Photonics, Scenario and Future Trends, Committee on Science and Technology of the Council of Europe (1983), BEPA 2 (1984), 37-41

2. A. Sona, Panoramica das Actividades na Europa, LASER PORTUGAL - 25 anos depois, Porto (1985)

3. O.D.D. Soares, L.M. Bernardo, Lasers - Princípios e Aplicações, AIP, A Industria do Norte 11 (1985), 52-73

   L.M. Bernardo, O.D.D. Soares, Processamento de Materiais com Lasers, AIP, A Industria do Norte 12 (1985), 30-39

4. O.D.D. Soares (Ed), Optical Metrology, Martinus Nijhoff, Dordrecht (1985)

5. L.A. Weaver, Machining and Welding Applications, in Laser Applications, Edt Monte Ross, Academic Press, NY (1974)

6. A.L. Schawlow, Lasers in Historical Perspective, IEEE Journ. QE-20 (1984), 558-561

7. E. De Rousse, Technology Update: Industrial Lasers, Optical Spectra, Sept (1981), 57-62

8. J. Steffen, Laser Materials Processing, in Lasers et Applications Industrielles, Presses Polytechniques Romandes, Lausanne (1982), 287-312

9. L. Holmes, Commercial Lasers - The Next Five Years, Laser Focus (May 1985), 146-154

10. F.P. Gagliano, R.M. Lumley and L.S. Watkins, Lasers in Industry, Proc IEEE 57 (1969), 395-428

11. Carslaw, H.S., Jaeger, J.C. Conduction of Heat in Solids, Oxford Univ. Press, London (1959)

12. A.S. Teixeira, Investimentos em Novas Tecnologias, LASER PORTUGAL, 25 anos depois, Porto (1985)

13. I.J. Spalding, Laser Applications, Phys Bull 35 (1984), 425-427

14. P. Harris, Laser Machining: Already Indispensable, Photonics Spectra, Oct (1983), 113-118

INDUSTRIAL APPLICATIONS OF LASERS OVERVIEW

by Alberto SONA - CISE - SEGRATE
Director of the Italian CNR Program on High Power Lasers

INTRODUCTION

Lasers applications in industrial materials processing can be subdivided into two major groups:
A- Applications requiring small but carefully controlled amount of energy such as micromachining ,resistor trimming,substrate scribing,semiconductor annealing,marking,printing etc.
B- Applications requiring substantial amounts of energy to induce the required phase transformation in the workpiece for processes such as cutting ,welding, heat treating and cladding.

Laser Efficiency and Power are not so important for processes A) whereas they are specially relevant for processes B).In both cases beam quality is a much desired feature with a different role in various processes.

Due to the less stringent requirements the lasers suitable for class A) processes are numerous and provide a broad choice of wavelengths which are often an added constraint for the specific application.
A non exhaustive list of lasers suitable for the first group of applications is the following:
Excimers;Ion (Argon - Krypton) ; Metallic Vapours (Cadmium - Selenium - Copper -Gold) Neodimium in YAG or in Glass ; Semiconductors ; Erbium in YAG ; Carbon Oxide ; Carbon Dioxide.

Efficiency and Power scalability requirements have restricted up to now the lasers usable for the second group of processes to Carbon Dioxide and Neodimium in YAG lasers .However not all the potentialities of Neodimium in Glass (especially in the slab configuration) and of the Carbon Oxide lasers (which have high efficiency and can be transmitted through low loss chalcogenide glass fibers) have been exploited and many researches are in progress on this line(1,2).In addition Excimers lasers seems to be very promising for the next future due to their high efficiency ,the relatively high average power levels (in the range of several hundreds watts) and the higher absorption by most metallic materials at these wavelengths.

On the short term however Carbon Dioxide and Neodimium in YAG lasers will be the workhorses for the large majority of class B) processess.

No industrial applications are expected on the other hand for chemical lasers such as HF or DF or for gasdynamic CO or CO2 lasers in spite of the extremely large amount of power they can deliver.Actually both of them are not closed cycle and in addition toxic chemicals are released by HF and DF lasers thus rendering industrial applications in a factory absolutely unpractical.

Soares, O.D.D., Perez-Amor, M. (eds), Applied Laser Tooling. ISBN-13: 978-94-010-8096-5
© *1987. Martinus Nijhoff Publishers, Dordrecht.*

LASERS IN MATERIAL PROCESSING

Materials processing and in particular metalworking have beeen implemented in the past centuries mainly by three classes of methods with different energy delivery procedures namely purely mechanical tooling, thermomechanical and purely thermal methods.

The first ones have in general higher accuracy; their action consists
in the breaking of lattice bonds by mechanical stresses and the concentration of the action derives from the tool geometry.When mechanical energy is
applied an unwanted deformation of the workpiece and tool wear results.To
allow for maximum accuracy the mechanical action has to be reduced in the
final processing stage together with the processing speed.

Thermomechanical methods such as hot forging or shaping by extrusion
consists in general of a mechanical action on materials softened by thermal
energy which in these cases is not concentrated.When only thermal energy is
used such as in casting the molten material derives its shape from the modeling frame with minimum mechanical action.

Material removal with no mechanical action can be accomplished by purely thermal methods such as flame or plasma cutting based on the delivery
of concentrated fluxes of thermal energy.The accuracy of thermal methods is
in general lower but they allow easier and faster materials processing due
to the larger amounts of energy involved.Metallurgical transformations are
also accomplished by purely thermal methods.

The laser beam is actually a new tool which can deliver thermal energy
with a concentration typical of mechanical methods and a similar accuracy.
In comparison with conventional processes the laser has the following features:
- it does not exert any mechanical action on the workpiece allowing simpler
fixturing,
- it has no inertia and can be easily moved,
- there is no wear or corrosion of the tooling edge,
- it allows typical processing of mechanical methods such as material removal for cutting and drilling,
- it allows processing typical of thermal methods such as welding ,surface
hardening and cladding,
- it can be easily adapted to different tasks by changing the focusing system and the related energy density at the target,
- due to the very fast and concentrated energy delivery a smaller amount of
material is involved in the process with reduced distorsion and narrower
heat affected zones at the sides .

Because of these properties it is the candidate most efficient as the
tool both for dedicated material processing and for flexible automated manufacturing systems.

The energy release to the workpiece occurs at a very high rate due to
the possibility of concentrating the laser beam in regions of the size of a

few wavelengths.A laser beam can be focused in a spot with a diameter d=2.44 λ f,where λ is the wavelength and f=F/D is the f-number that is the ratio between the focal length F and the beam diameter D.As an example a 1 kW CO2 laser beam can be concentrated by a lens providing an f-number =8 in an area of about.03 mm square providing a power density of 3 Mw/cm square.- The induced thermal field is characterized by extremely high temperature gradients and time derivatives in comparison with conventional energy delivery methods.With short interaction times the heat diffuses at small distances .The diffusion length L is given by L = 4D t where D is the thermal diffusivity and t the interaction time (D = 5 mm square/s for iron and ferrous materials).As an example the diffusion length in iron in a time t=0.o5s is one mm.

Only the electron beam has an action on the workpiece similar to the laser .Electron beam processing has to be performed under vacuum and only on metallic materials.Laser beams are of more practical use ; however they cannot reach ,at least in air, the same penetration in metals as the electron beams.

INDUSTRIAL APPLICATIONS

The laser processes can be divided into two major groups namely:
-High energy density processes requiring the melting of the material in the interaction region such as Cutting ,Drilling and Welding accomplished by focused beams.
-Medium energy density processes requiring only the softening or the heating just close to the melting temperature such as Cladding,Surface Hardening,Annealing,Glazing.

In Table I the power densities and the related interaction times are reported for the various processes.

A summary of the working principles of the various processes is given in the following :

1 - Cutting ,Drilling or Material Removal.
Two basic processes are possible:
a) - The laser beam melts the material and the removal is accomplished by a gas jet.
b) - The laser beam vaporizes the material and removal occurs by the expanding vapour .
The first process, referred as melt-cutting is typical of most CW beam cutting systems.The second usually occurs with pulsed beams having a power density of an order of magnitude higher (100 -1000 watts /cm square).

2 - Welding or Joining.
Again two basic processes are possible:
a) - Conduction welding where the energy propagates from the surface to the bulk material.This is typical of low power density processing and results in large molten regions and heat affected zones.
b) - Deep penetration welding where the energy is released inside the "key

hole"which is formed under the action of the high pressure vapour coming from the molten region.A heat line source normal to the surface is formed inside the material down to the penetration depth.This second mechanism is more effective and leads to narrower welds and heat affected zones (fig.1).

3 - Surface Hardening by Heat Treatment.It can be obtained:

a) - By heating the surface at a temperature below the melting point and by allowing for a fast cooling rate after the laser beam removal resulting in a self hardening of the transformed region.

b) - By heating the surface above the melting point and allowing for its resolidification followed by self hardening.

The first method does not impair the geometry of the workpiece ;the second one requires the subsequent reworking of the surface.

4 -Cladding and Hardfacing.

It is based on the softening of the surface layer to allow the inclusion of an external material f.i. wear resistant.

The deposition can be:

a) - Static if the powder is pre-placed locally with a suitable binder.

b) - Dynamic if the powder is injected by transport gas jets in the laser softened region.

5 -Surface Alloying and Glazing.

It is a process similar to static cladding but with special control of the composition and of the cooling rate to allow for the formation of alloys with special metallurgical structures.

6 -Laser Assisted Machining.

Different processes have been studied :

a) - Mechanical removal by conventional tools of laser softened materials ,

b) - Softening by a linearly focused beam of the region where the mechanical displacement is induced by the cutting tool.

c) - Lathe -like material removal by simultaneous radial and longitudinal cuts performed by two focused beams at 90 degree resulting in the detachment of annular chunks of material.

7 - Reverse Machining .

The process consists in cladding with a powder made of the same material of the piece to be treated.This allows restoring the geometry of wear eroded components such as axles near the supporting bearing or turbine blades.A subsequent conventional machining can restore the geometry of the component ; this procedure was used to repair expensive turbine rotors.

Processes 1 to 4 are already in use in manufacturing plants,processes 5 to 7 are still under investigation.

In table II some examples of components manufactured using laser are reported for the different processes.

In table III a list of the industrial activities where lasers are now being used is given together with possible new areas of interest (within brackets).

The major advantages of laser processing in industry are:
- Highly increased productivity (X 3 - X30)
- Significant reduction in materials and processing costs(20-30 %)
- Reduction in manpower.
- Higher product quality kept within narrow limits.
- Reduction / elimination of scraps.
- Improved conditions of the workshop environment (reduced noise,vibrations,fumes,chips, lubricating fluids)
-Capability of performing the metallurgical treatments on the stations of a mechanical line.
-Production systems easy to be controlled and automated with very fast response times.
-Several kind of processes can be performed on different materials with no change of tooling.
-Production systems ideally suited for incorporation in flexible production lines.
-Dedicated systems with full numerical control of the process (feedback loops can be arranged with suitable sensors)

The major disadvantages limiting at present the diffusion of laser processing are:
- the cost of the equipment which limits the application to high added value products,
- the lack of diffused information on the laser processing technology,
- the need of redesign the mechanical parts and review the manufacturing procedure to exploit at best the laser technology.
All the above limiting factors will be decreasing in importance in the future.
Arguments against lasers expressed in the past such as :
- Difficult to install
- Requires too much maintenance
- Costly to operate and to maintain
are no longer valid.

INDUSTRIAL SYSTEMS AND BEAM DELIVERY PROBLEMS

The requirement of integrating a laser in a Flexible Manufacturing System or in Robot Assisted Laser Machining poses the problem of beam delivery to the workpiece which has in general five degrees of freedom with respect to the beam focusing head.In addition a sixth one has to be considered for the auto focus feature.Depending on the mass and size of the workpiece the five degrees of freedom can be shared in different ways between the laser focusing head and the worpiece itself(fig.2). The simpler beam delivery system is ,of course,the one required when the workpiece is given all the degrees of freedom.The opposite case is the one ,typical of robotic

systems, where the workpiece is fixed and all the five degrees of freedom
are given to the focusing head thus implementing a five axis machine.The
robotic system can be polar or of the gantry type ;in any case it has to
provide relative positioning with an accuracy of the order of the focal
spot size (typically of 100 microns) which is not usual for conventional
robotic tooling systems.

In general a simple solution can be provided by positioning with a robot the beam focusing head which has to be connected to the laser head by a
suitable flexible delivery system namely an articulated arm (fig.3)or an o-
ptical fiber.This solution allows the use of already existing robots ,pro-
vided they have the necessary accuracy.However it requires a rather sophisticated software able to avoid mechanical interferences which can damage
the flexible optical link.In addition more or less severe limitations to
the kinematics of the robot are introduced with both arrangements.   Neodymium in YAG laser beams with an average power up to 600 watts have been
conveyed in a fused silica optical fiber with a 1.2 mm core diameter and 15
meters length.The losses in the system were smaller than 10 percent and were mainly due to the end faces.Smaller core diameters could be used at lower power levels,the limit being set by the beam divergence (15).Under this
respect a substantial improvement is expected by lasers with slab geometries providing low divergence beams due to optical index gradients compen-
sation.Research along this line is in progress in various industrial la-
boratories (2,16).

On the long term the best solution will be the use of specially designed hollow robots with a inner free optical path providing flexible optical guiding without added kinematic constraints and suitable positioning
accuracy.

Flexible Manufacturing System will require also appropriate beam delivery systems for feeding the different workstations.Two different approaches can be considered depending on the specific manufacturing processes
(17).

One possible approach makes use of a single very high power laser unit
delivering the beam to different workstations on a time sharing basis.All
the power can be conveyed on a single workstation at a time or more
workstations can be fed simultaneously with a fraction of the total power
(fig.4).

A second possible approach consists in using a few smaller power laser
beams which can be combined by suitable optical systems providing a single
beam with a total power equal to the sum of the individual ones or two or
more beams having a power resulting from the superposition of the selected
beams.The generated beams can then be forwarded to the different workstations with the appropriate power level selected according to the process
requirement (fig.5,6).The combining-switching matrix can be implemented by
mobile computer controlled optical elements (7).

The main advantages of the second approach is the use of smaller power units ,in general more reliable,by the possibility of adding further lasers in a modular fashion ,by the substantial reduction of the risk of a comple-te stop because of the multiple laser heads in comparison with the case of the single high power laser going out of service.Finally a back-up unit is not as expensive.The disadvantages are the major investment cost for the same total power installed.

From the application point of view the first approach is more appro-priate when most manufacturing processes do require the full power whereas the second is more useful when most processes require a fraction of the to-tal power and only a few need the total.

CONCLUSIONS

A continuous growth of high power lasers applications in industry is expected.On one side the diffusion of the laser manufacturing technology for low added value products will require low cost dedicated laser sy-stems.For mass production and /or for high added value components high per-formances such as good beam quality ,reliability , computer controlled e-mission are required to allow fully automated processing.The added require-ments set by the use of lasers in connection with robots and flexible manu-facturing systems need further steps in industrial lasers technology.Sophi-sticated technologies for generating high power high quality laser beams with accurate delivery systems are already available for other applica-tions.Their use in industry is limited by economical problems which will become less severe in time as the use of laser will be further expanded.A key point to speed up this process is the development of standard modular subsystems for high power laser beams generation and delivery.This would result in a cost reduction and in reliable and controlled performances as it occurred already in the machine tooling industry.Possible future appli-cations are expected in nuclear and aerospace industries in the short term and on a longer time scale in shipbuilding and iron industries when higher power laser units will be available for practical uses. The market of laser systems for material processing is expected to reach in the western world the 300 millions of dollars level in 1985 with an increase of 27 % over the 1984 (8).A similar annual growth rate is expected for the nexth five years.

REFERENCES

1) - H.Saito.t al.   Proc. CLEO '85,paper WM30,pg.120.

2) - W.B.Jones Laser Focus / Electro Optics Sept. '83.

3) - G.H.Sugawara et al. Proc. CLEO '84 , paper TUC3,pg.54.

4) - A.V.La Rocca et al. Proc. LASEROBOTICS I 1985 ,paper No.4.

5) - M.G. Jones et al. Proc. ICALEO '83 ,pg.148.

6) - D.Plankenhorn et al. Proc. LASEROBOTICS I 1985 ,paper No.15.

7) - V.Fantini et al. Proc.Laser in Manufacturing '85 ,pg.249.

8)   - C.Breck Hitz The Marketplace for Material - Processing Lasers
Lasers and Applications Jan.85 pg.5

GENERAL REFERENCES

A)   - H.Koebner INDUSTRIAL APPLICATIONS OF LASERS ,JOHN WILEY AND SONS
Publ. 1984
B) - D.Schuocker Proc. of the Meeting on INDUSTRIAL APPLICATIONS OF HIGH
POWER LASERS Sept.26 - 27 ,1983 Linz,Austria SPIE Vol.455
C) - A.V.La Rocca Laser Applications In Manufacturing Proc. of the Confe-
rence on THE FACTORY OF THE FUTURE Sept 83 Stockholm IVA Rapport 258
D) - M.F.Kimmitt Proc.of the Conference on LASERS IN MANUFACTURING 26 -28
March 85 Birmingham U.K.
E) - E.A.Metzbower Proc. of the MATERIALS PROCESSING SYMPOSIUM ICALEO 83
Vol.38 Publ.by the L.I.A.
F)   - Conrad M. Banas "Macro Materials Processing " Proc. of the IEEE
vol.70 June 1982
G)   - John F.Ready "Material Processing - An Overview" Proc. of the IEEE
vol. 70 June 1982.

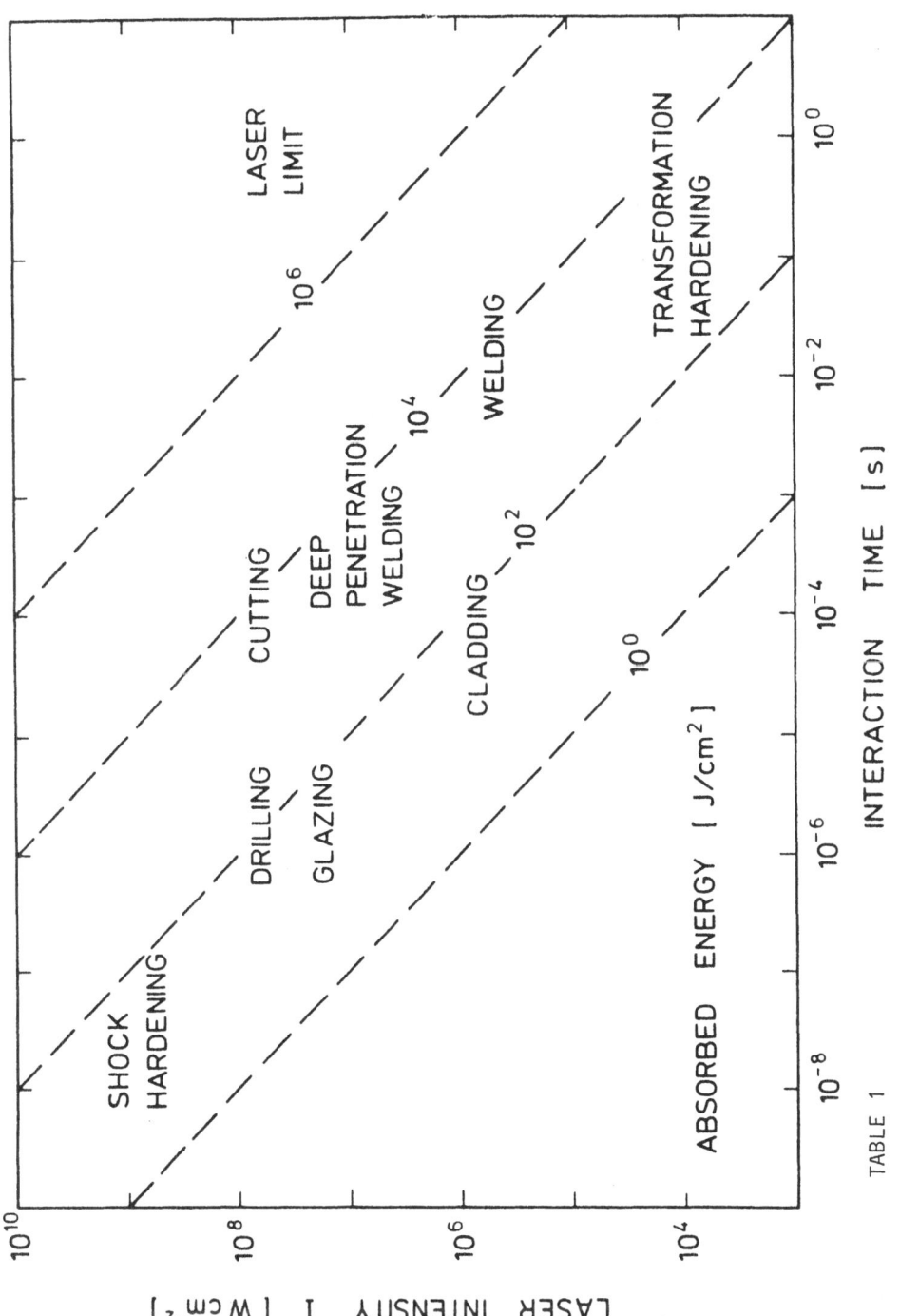

TABLE 1

TABLE II

EXAMPLES OF COMPONENTS PROCESSED BY LASERS

CUTTING
- Iron sheets for various applications
- Deeply stamped iron sheets for car bodies
- Carpets on preformed plastic support for cars floor
- Textiles for dresses
- Leather for shoes and gloves
- Plastic sheets for signs
- Plastic laminates for eyeglasses mounts
- Ceramic substrates for microelectronics

DRILLING
- Holes in turbine blades to control the gas flow
- Holes in filter cigarettes to allow dilution of smoke by air
- Holes in baby nipples to allow regular flow of the liquid

WELDING
- Steel plates with different geometries
- Iron coils for continuous sheet rolling mills
- Continuous welds for stainless steel pipes
- Synchro gears for car shifts
- Lead electrodes for long life batteries
- Copper - titanium electrodes for electrochemical cells-
- Aluminum spacers for double glasses windows

CLADDING
- Exhaust valves and seats for car engines with stellite
- Turbine blades with anti-wear alloys

SURFACE HARDENING
- Inner bores of Diesel engines
- Automatic shifts gearboxes to increae wear resistance
- Various shift leverages

TABLE III

INDUSTRIES USING LASERS IN MANUFACTURING

- AUTOMOTIVE INDUSTRIES: CAR BODIES AND ENGINES

- ENERGY AND ELECTROTECHNICAL INDUSTRIES:HEAT EXCHANGERS ,TURBINES,MOTORS

- ELECTRONIC INDUSTRY: RESISTOR TRIMMING,SUBSTRATE SCRIBING, MARKING

- AEROSPACE INDUSTRIES: TITANIUM AND ALUMINUM ALLOYS,COMPOSITE MATERIALS

- GENERAL MECHANICAL INDUSTRY :   PRECISION CUTTING ,WELDING,HARDENING

- GLASS INDUSTRIES: CUTTING OF GLASS PLATES, DRILLING.

- IRON INDUSTRIES: CUTTING AND WELDING OF IRON SHEETS.

- LIGHT MANUFACTURING INDUSTRIES: TEXTILES,WOOD,PLASTICS,LEATHER.

- (NUCLEAR INDUSTRIES: FUEL RODS REPROCESSING,SEALING FUEL CONTAINERS.)

- (SHIPBUILDING INDUSTRIES: CUTTING AND JOINING IRON PLATES)

36

FIG. 1   THE TWO TYPES OF LASER WELDING

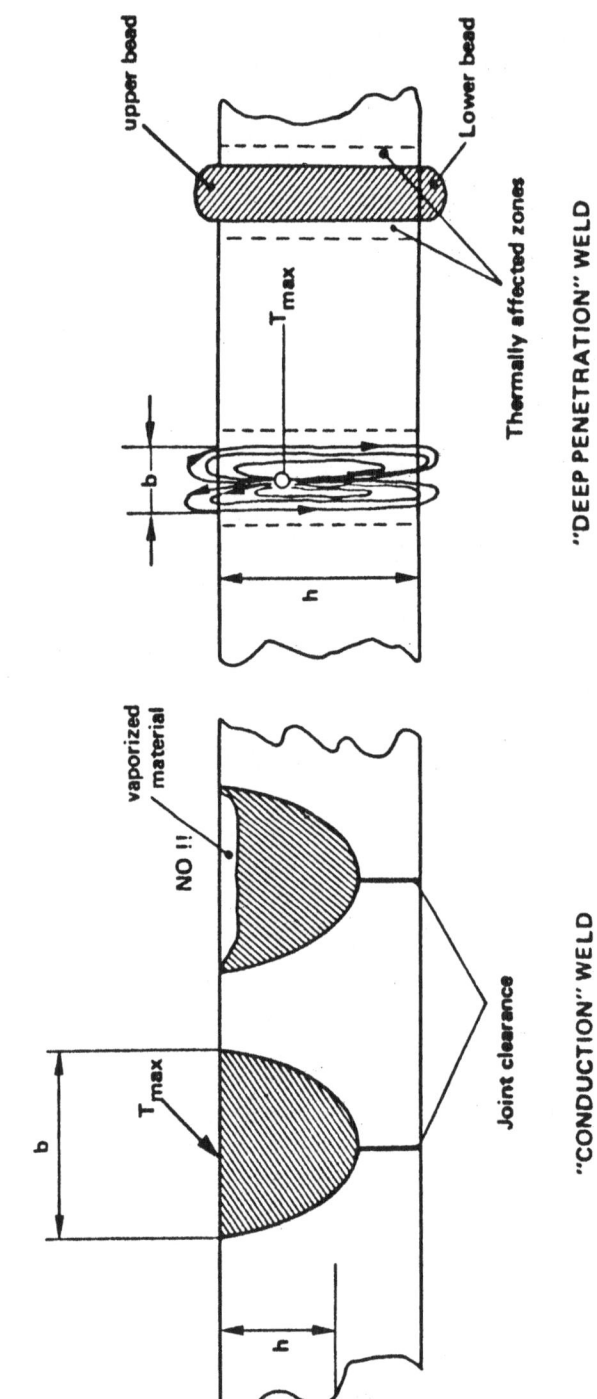

h/b = 1→1.5

h/b = 6→10(12)

"CONDUCTION" WELD

"DEEP PENETRATION" WELD

37

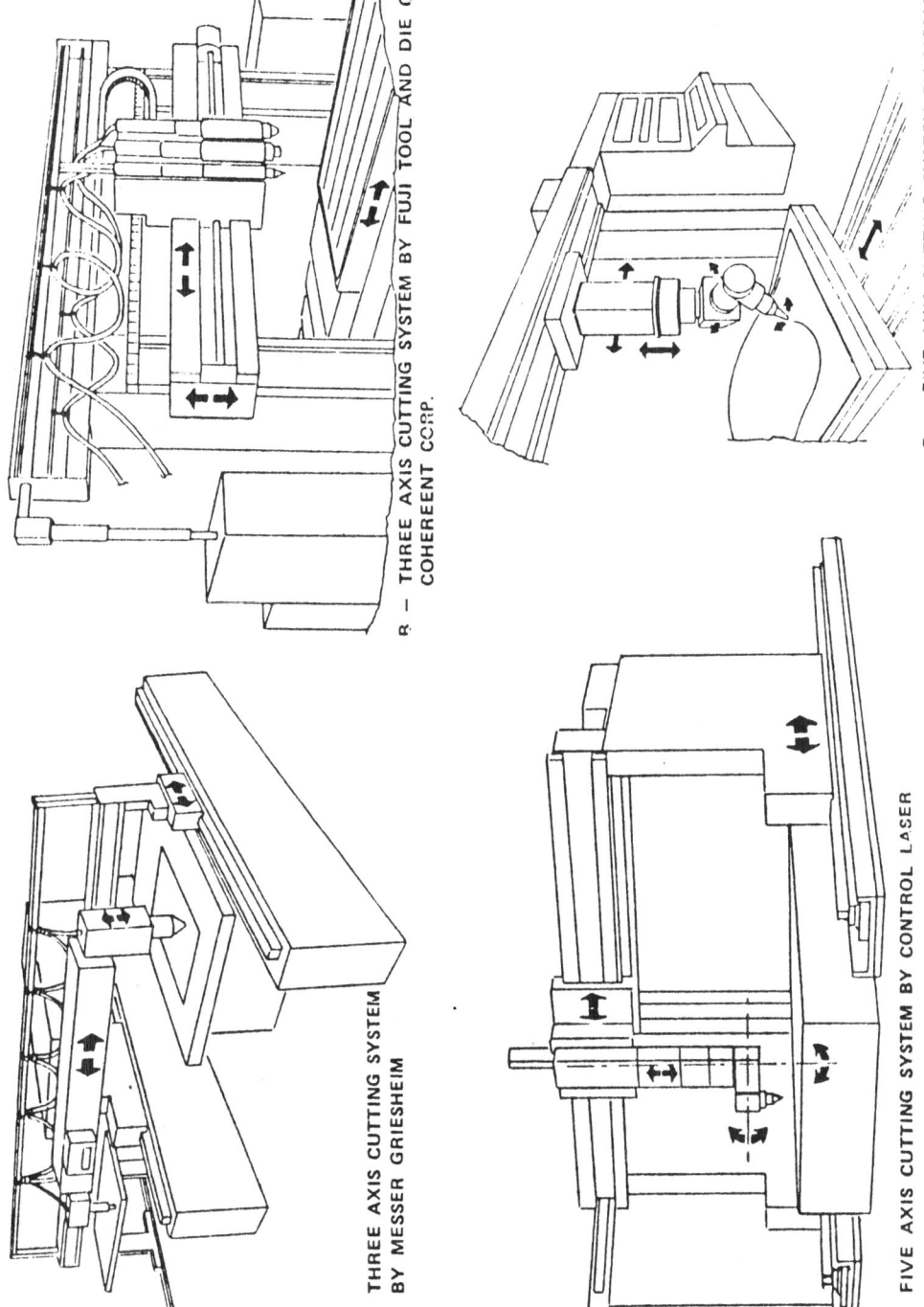

A — THREE AXIS CUTTING SYSTEM BY MESSER GRIESHEIM

B — THREE AXIS CUTTING SYSTEM BY FUJI TOOL AND DIE CO./ COHEREENT CORP.

C — FIVE AXIS CUTTING SYSTEM BY CONTROL LASER

D — FIVE AXIS CUTTING SYSTEM BY MESSER GRIESHEIM

FIG. 2 — GANTRY SYSTEM CONFIGURATIONS

38

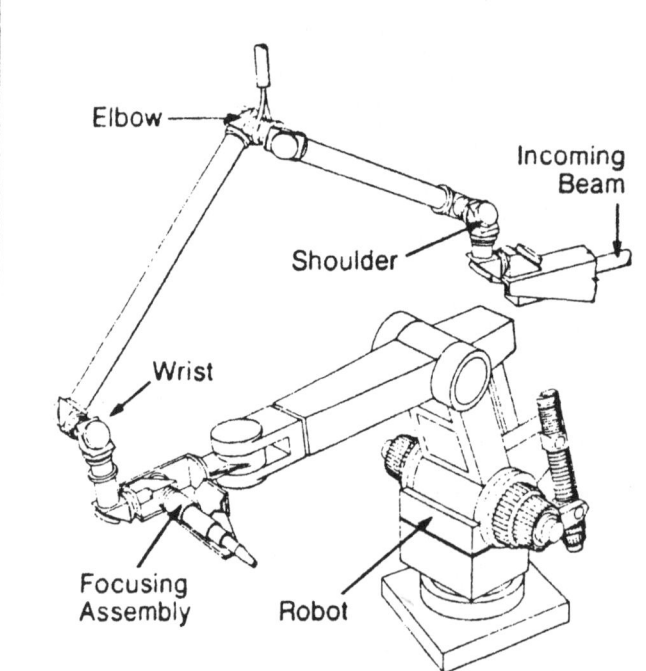

Fig. 3a A typical floor mounted configuration

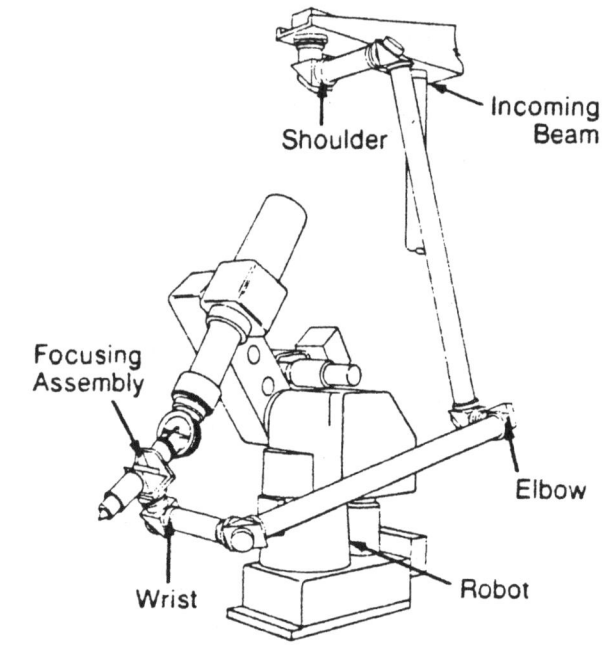

Fig. 3b Laser-robot with overhead mounting

LASER CENTER

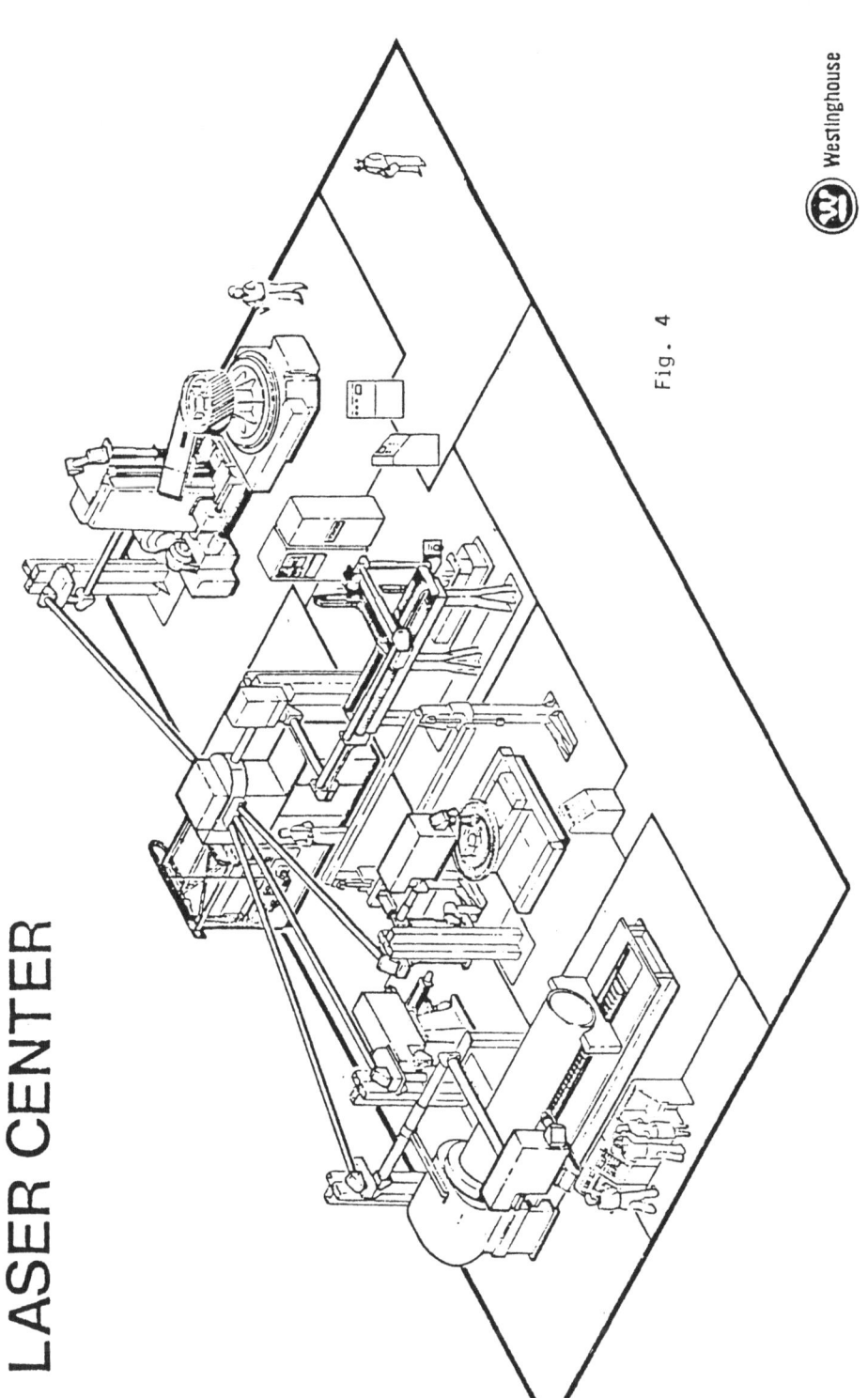

Fig. 4

INDUSTRIAL HIGH POWER LASER PROCESSING FACILITY,
WITH 6 WORKSTATIONS

(W) Westinghouse

40

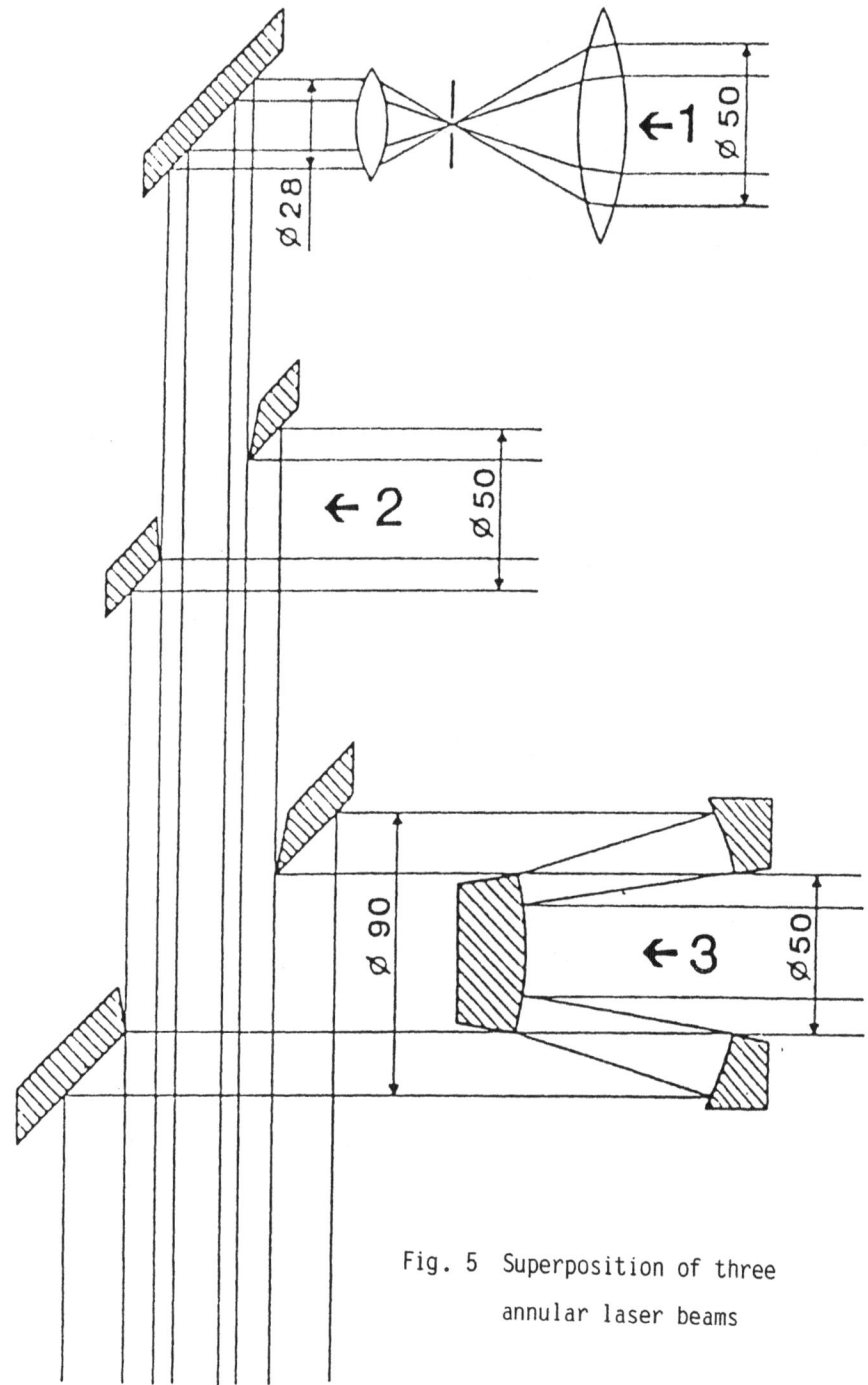

Fig. 5 Superposition of three
annular laser beams

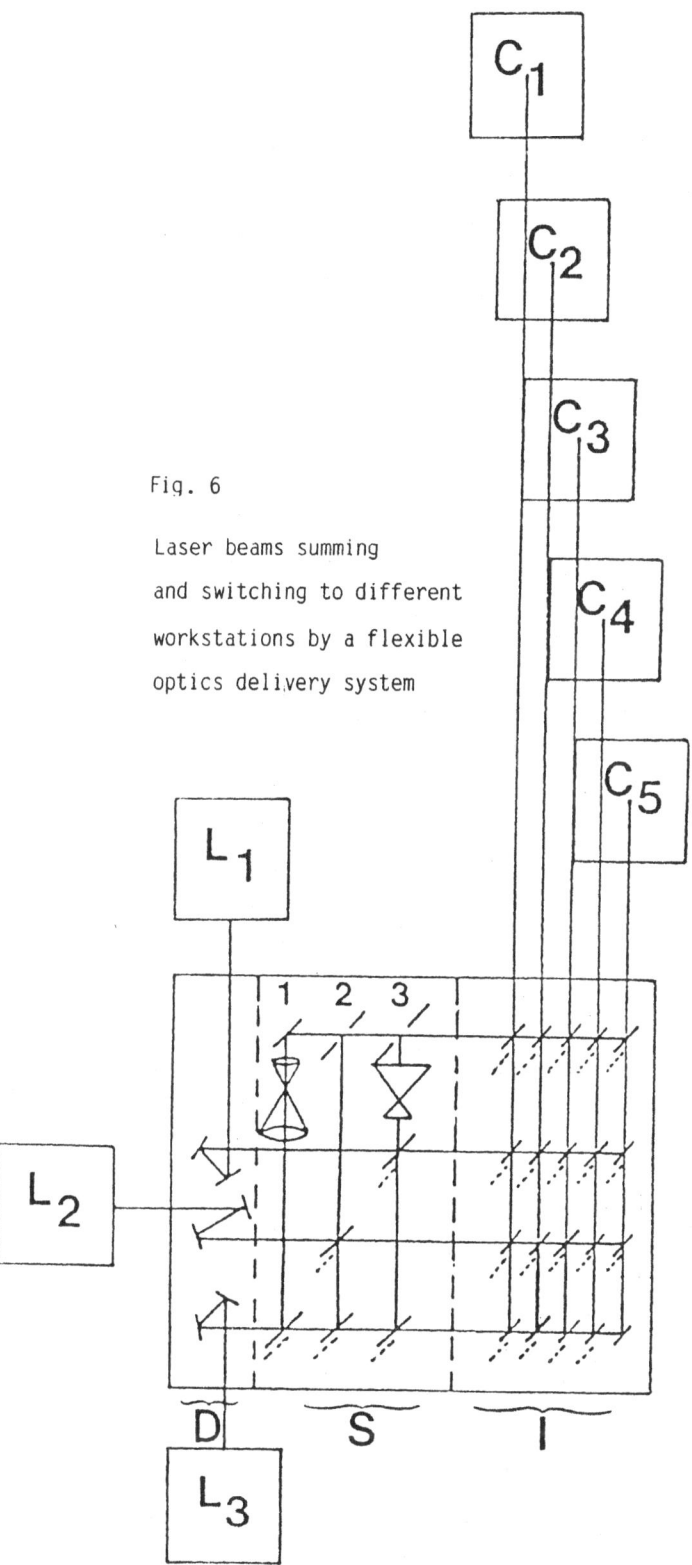

Fig. 6

Laser beams summing
and switching to different
workstations by a flexible
optics delivery system

THE PHYSICS OF LASERS

B.M. LEON-FONG

E.T.S. de Ingenieros de Telecomunicación
Universidad de Santiago
Vigo, SPAIN

1. INTRODUCTION

The principles of lasers are briefly reviewed for readers concerned with laser applications as an industry tool, not requiring a specialized knowledge of lasers. References are given for more in-depth information.

LASER is an acronym for Light Amplification by Stimulated Emission of Radiation coined by T.H. Maiman (Ref.1) for his newly invented device in analogy to "maser", which works at microwave frequencies instead of at light frequencies. Maiman in 1960 succeded in overcoming all difficulties and constructed the first laser (a ruby one), just two years after the theoretical work of Schawlow and Townes 1958 (Ref.1) on the extension of the maser principle to the optical region. In the few months thereafter several research groups were able to develop other laser systems & during these 25 years of laser existence a real avalanche of research and applications on and with lasers has taken place. Moreover we count today with a wide variety of efficient and reliable commercial lasers, as reviewed in Ref.8.

2. PARTS OF A LASER

As its name states, the laser is a light amplificator, and as such every laser consists of three main parts:
   a) the amplifying or active medium,
   b) the pumping source and
   c) the resonator or optical cavity, which provides for feedback.

2.1. The active medium

Atoms and light can interact mutually. How does this take place? When an atom is exposed to an electromagnetic wave, in which the electrical field is oscillating, its charge distribution will be periodically disturbed by the field. This perturbations can be understood as a composition of the different allowed charge distributions, prescribed by quantum mechanics. But quantum mechanics make statistical statements, thus due to the presence of an electric field, we get certain probabilities to find atoms in higher excited states. The transition energy has been taken from the field -light-, and an absorption process has ocurred. As most basic processes, this mechanism is reversible, i.e. by emission of light, the excited atom give energy to the electromagnetic field and goes back to the ground state.

   2.1.1. Radiative energy exchange. As an introduction to ener-

Soares, O.D.D., Perez-Amor, M. (eds), Applied Laser Tooling. ISBN-13: 978-94-010-8096-5
© *1987. Martinus Nijhoff Publishers, Dordrecht.*

gy exchange processes by radiation, the basis of laser action, we shall follow Albert Einstein's arguments (Ref.2).
We consider an atom or a molecule with quantised states

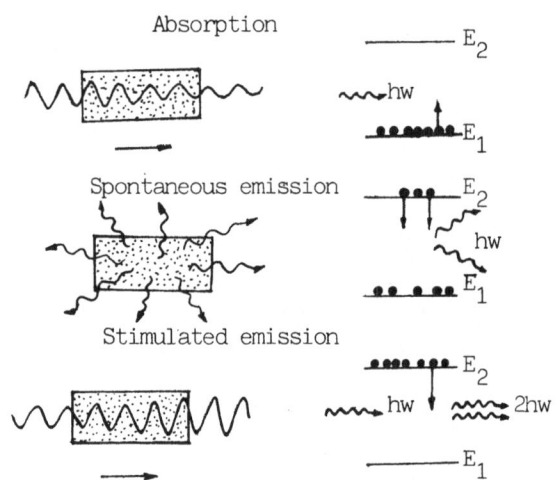

FIGURE 1. Energy exchange processes by radiation (Ref.4).

1 and 2 (Fig.1) with energy states E_1 and E_2 ($E_2 > E_1$) respectively. Radiative transfer between the two energy states be allowed. The molecule can transfer from state 2 to state 1 by emitting energy, or conversely from state 1 to state 2 by absorbing energy. The energy removed from or added to the molecule or atom appears as energy quantum hw

$$\hbar w = E_2 - E_1 \qquad (1)$$

Without external stimulation, an excited atom in the higher state 2 returns spontaneously to the lower state after an average time τ. This is the "spontaneous emission". If A_{21} is the transition probability from state 2 to 1, also called the Einstein coefficient for this process, then the probability to get spontaneous emission in a certain time dt is

$$d W_s = A_{21}\, dt \qquad (2)$$

Since spontaneous emission is a statistical function of space and time, having a large number of spontaneously emitting molecules, there is no phase relationship between them, so that the emitted light quanta won't be coherent. If there are N_o excited (E_2) molecules at time t=0, then the number of excited molecules decreases exponentially with time as

$$N = N_o\, e^{-t/\tau} \qquad (3)$$

being the time constant τ equal to the reciprocal of the transition probability

$$\tau = 1 \ / \ A_{21} \tag{4}$$

The analogous classical concept to spontaneous emission would be the radiation of a damped harmonic oscillator.

Now, if the molecules are in a radiation field of energy density $\rho(w)$, according to Einstein, the probability to induce transitions by the field will be proportional to the energy density. The field can either supply energy to the molecule or atom or be amplified by the emitted photons from the molecule or atom. The first process would correspond to an "induced absorption", if the molecules were originally in the lower energy state 1. The absorption probability would be

$$d \ W_{12} = B_{12} \cdot \rho(w) \cdot dt \tag{5}$$

B_{12} and B_{21} are the transition probabilities for processes 1 to 2 and 2 to 1 respectively.

The second process corresponds to a "stimulated emission", in which the molecule gives up a photon to the radiation field, if it was previously excited in state 2, the probability for this process being

$$d \ W_{21} = B_{21} \cdot \rho(w) \cdot dt \tag{6}$$

While the probability of induced transition is proportional to the energy density of the external radiation, spontaneous emission is independent of external fields. There is another important point, which will affect the properties of laser radiation: in contrast to spontaneous emission, in the case of induced transition, there is a fixed phase relationship between the stimulating field and the emitted radiation. They are coherent.

At temperature T in thermal equilibrium, the population numbers N_2 and N_1 of the states 2 and 1 result from the Boltzmann distribution

$$N_2 \ / \ N_1 = e^{-(E_2-E_1)/kT} = e^{-\hbar w/kT} \tag{7}$$

It can be shown (Ref.3-4) that the Einstein coefficients for induced processes are equal: $B_{12}=B_{21}=B$.

2.1.2. Energy balance for radiation. Considering radiation of energy density $\rho_o(w)$ incident on a substance as a near plane wave, after the radiation has penetrated a distance x into the material, its energy density will be $\rho(w,x)$. The radiation excites both absorption as well as stimulated emission. The decrease in energy density due to absorption between states 1 and 2 is

$$\left(\frac{d \ \rho \ (w,x)}{dt} \right)_{abs.} = - \ N_1 \ \rho(w,x) \ g(w) \ B \ \hbar w_o \tag{8}$$

where $g(w)$ is the frequency dependency of the transition probability, which has a sharp maximum at w_o, and N_1 is the number of atoms available for absorption.

The transition from state 2 to state 1 by stimulated emission increases the energy density of radiation by

$$\left(\frac{d \ \rho \ (w,x)}{dt} \right)_{St.Em.} = + \ N_2 \ \rho(w,x) \ g(w) \ B \ \hbar w_o \tag{9}$$

During the time dt the radiation penetrates a distance dx in the active material, whose refractive index is n:

$$d \; x = c_o \; / \; n \quad dt \; = c \; dt \tag{10}$$

where c and c_o are the velocities of light in the medium and in vacuum.

Neglecting the spontaneous emission as a source of noise, the total change of energy density, taking into account absorption and stimulated emission as well, is

$$\frac{d \; \rho \; (w,x)}{dt} = (\; N_2 - N_1 \;) \; \rho(w,x) \; g(w) \; \hbar w_o \; \frac{1}{c} \tag{11}$$

From this equation we can observe three main interaction cases between light and matter:

i) $N_1 > N_2$: The population of the lower level is larger than that of the upper level. The radiation is attenuated as it passes through the material ($d\rho/dx$ <0). In thermal equilibrium (equation 7), radiation attenuation is always observed, since the population of the higher energy level is always smaller than that of the lower level.

ii) $N_2 = N_1$: ($d\rho/dx = 0$) The radiation passes through the material without any change. The material is transparent to the incident radiation.

iii) $N_2 > N_1$: ($d\rho/dx > 0$) This state is called <u>population inversion</u> and corresponds to a non-thermal equilibrium. The point at which the population of both states is equal is called the "inversion threshold". Once achieved this population inversion, the energy density increases as the radiation passes through the material.

In order to produce population inversion, we must have, in addition to black body radiation (which produces a thermal distribution), a further source of energy to populate preferentially a specified energy level. This is the pumping source.

2.1.3. <u>Inversion methods</u>. With the use of just two levels 1 and 2, it is impossible to produce population inversion in steady state. Mostly three- or four- level schemes are used. In a three-level laser (Fig.2a) the atoms are raised from the ground level 1 to level 3 by some pumping mechanism. If the material is chosen such that, after an atom has been raised to level 3, it decays rapidly to level 2, then in this way

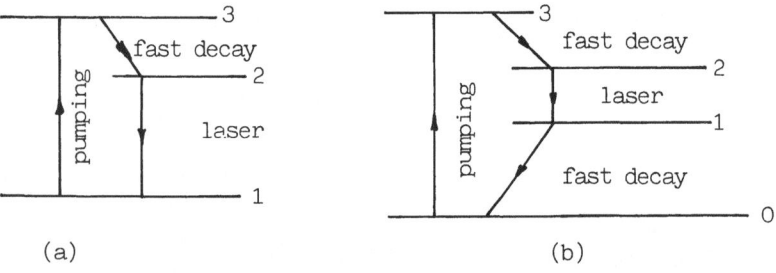

(a)   (b)

FIGURE 2. a) Three-level laser scheme. b) Four-level scheme.

a population inversion can be obtained between levels 2 and 1. The energy differences between the various levels are usually much greater than kT. According to Boltzmann statistics (Eq.7), one can say that at equilibrium essentially all N_t atoms are in the ground state (level 1), being N_t the total number of atoms per unit volume of material. If we now raise atoms from level 1 to 3, they will decay to level 2 very fast, so that level 3 will remain more or less empty. This means that we have to raise at least half of the total population to level 2 in order to achieve condition ii). It is only from this situation on, that we can produce population inversion by raising more atoms.

We still can get population inversion more easily, if we work with a four-level scheme (Fig.2b). If transitions 3 to 2 and 1 to 0 are very fast, from the very beginning on, level one is empty, and every atom raised to level 3 will decay rapidly to level 2, contributing thus to population inversion.

Fig. 3 shows the laser energy schemes for ruby and neodym with the corresponding decay times, as examples for three- and four-level systems.

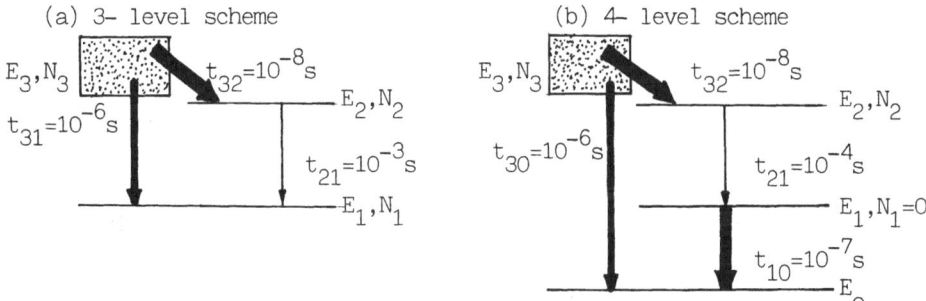

(a) 3- level scheme

(b) 4- level scheme

FIGURE 3. a) Ruby laser scheme. b) Neodymium laser scheme.

There are many types of active media, each giving the name to the particular laser. From all different types of lasers we will only review very briefly those more interesting for material processing at the end of this chapter.

2.2 The pumping source.

According to the different energy states schemes, population inversion requires excitation of atoms or molecules to the higher laser level by other methods, called pumping methods. Since there are now a multiplicity of currently known laser transitions and a great complexity of each individual system, we will only mention here some of the mechanisms.

2.2.1. Optical pumping. The pump energy is supplied by an intense light source with three- or four-level schemes. This pumping method is possible with gases, liquids, dielectric solids and semiconductors. In gases at low pressure all transitions are narrow band transitions, thus the pump source must be a narrow band one, with its frequency coinciding sufficiently closely, within a small part of kT, with the frequency of a pump transition in the gas to be inverted (Example: caesium laser with a helium tube lamp as pump source).

This method is also particularly important with dielectric

crystals, because this is the only possibility of introducing sufficient energy into such substances. In most cases, the active material is pumped according to a three- or four-level scheme, but it is also possible to pump into broad bands of an auxiliary material, which then gives up its excitation energy by resonant exchange selectively to the laser atoms. Besides a sharp fluorescent emission line, a laser substance suitable for optical pumping should have a broad band pump transition. Normally, broad band light sources are used as pump sources for optically pumped solids or liquids, such as incandescent lamp, xenon arc lamps, high pressure mercury lamps or xenon flash tubes. In some cases, as in dye lasers, it is very useful to use another laser as pumping source.

2.2.2. Gas discharge pumping. Energy can be supplied to the molecules or atoms of a gas by collision excitation. By this method one can achieve to populate a certain energy level in different ways, such as a) electron collision, b) dissociation of polyatomic molecules through collisions with excited atoms, c) direct selective excitation of an atomic species by collision with another excited atom with coinciding energy levels or d) combination of several processes. In many cases, processes taking place in the gas discharge are also useful to population inversion by depopulating the lower laser level (Ref. 3).

2.2.3. Current flow in semiconductors lasers. Inversion is achieved in the pn-junction of highly doped "degenerate" semiconductors, as gallium arsenide n-doped with selenium, tellurium, germanium or silicon to a carrier density of $3.10^{17} - 6.10^{18}$ carriers/cm^3. The p-layer is doped with zinc. The emission of light is produced by radiative recombination of electrons and holes in the pn-junction. They work on a four-level scheme if the valence band in the upper region is left empty and the conduction band is filled only in the vecinity of the lower boundary.

2.3. The optical resonator.
Once we have an active medium, which we have inverted by some pumping method, we can amplify light, but this isn't a laser yet. In order to have a laser, we have to provide feedback by placing the active medium in an optical resonator.

The optical resonator consists of two mirrors. One of them is totally reflecting, the other is partially reflecting, so that a proportion of the incident light is transmitted. The mirrors are usually aligned perpendicularly to the optical axis of the active medium, so that only light on the laser axis is reflected and transmited, resulting in the preferential build up of radiation along the direction of the laser light. This results in positive feedback, a rapid cumulative build up of radiation along the laser axis ocurring.

Because feedback only occurs along the axis of the laser, the light radiated in any direction different from the laser optical axis runs out of the oscillator and the remaining radiation is all directed along the laser axis as a narrow beam of low divergence governed by laser optics and geometry and limited by diffraction at the output mirror.

The length of the cavity selects output wavelengths by the

standing wave criteria, i.e. the cavity length must be an integer multiple of one half of the wavelength.

Another important function of the optical resonator consists of making longer the photon dwell time in the laser medium. Since the probability for stimulated emission increases proportional to the photon dwell time, laser process will be enhanced by the resonator.

Varios geometrical considerations including the radii of curvature of the mirrors and the length and diameter of the cavity determine the degree of feedback and whether laser action can occur. The conditions determining whether a geometrical ray in the cavity will be reflected in terms of the geometry of the cavity is given by

$$0 < (1 - d/R_1) (1 - d/R_2) < 1 \qquad (12)$$

where d is the length of the cavity and R_1 and R_2 are the radii of curvature of the mirrors. Ref.9 shows the stability conditions derived from Eq.12 for the parameters d/R_1 and d/R_2.

2.3.1. Resonator losses. In an optical resonator light reflects many times back and forth between the mirrors. If light falls perpendicular to the mirror, after each reflection some part of the incident power gets lost. The reflection loss for each reflection will be

$$\delta_R = 1 - R$$

R being the reflectivity of the mirror. Due to the reflections, light stays longer in the resonator volume. The dwell time for light in the optical resonator T is

$$T = \frac{L}{c} \frac{1}{1 - R} \qquad (13)$$

where L is the length of the resonator and c the light velocity in the medium. According to Eq.13, if we make R big enough, reflection losses can be made as small as wanted, so that we can achieve very high amplification by increasing the photon dwell time in the resonator. Diffraction and alignment losses limit this amplification though.

3. CHARACTERISTICS OF LASER

As laser light possesses a variety of properties not previously achievable by any other light source, their characteristics have been extensively studied in many books (Ref.5). Let us review those, that are most important for applications:
   a) Coherence,
   b) divergence,
   c) energy and power output and
   d) mode structure.

3.1. Coherence.

This is the most important characteristic of laser light. Some of the others are just a result of coherence. To explain coherence, the wave nature of light must be invoked. Radiation is said to be spatially and temporally coherent if the phase and amplitude of the wave at any particular time and position can be calculated from earlier known values. This is not the case of natural light sources as incancescent or gas discharge lamps, since there, light emission takes place spontaneously at each atom independently and at any time. Thus, there is no phase relation among the different emission processes.

Emission processes are also perturbed by Doppler effect and collisions, so that the best low pressure spectral lamps have "monochromatic" light with line widths greater than 10^9 Hz. This is equivalent to say that each atom has been performing undisturbed damped oscillations during an average time of $T=2.10^{-13}$ s. The length of the wave train would be $L_c = c.T_c =$ $=6$ cm. T_c is called the coherence time and L_c the coherence length.

Coherence is often called as the property of waves, which allows them to interfere. According to this, both temporal and spatial coherence are measured through interference experiments.

3.2. Divergence.

Conventional light sources emit in all directions. Consequently, intensity decreases with the inverse square of the distance. Refracting or reflecting collimators are only partly successful. Another source of divergence is diffraction ocurring at the boundaries of apertures.

In lasers, due to the resonator properties, only light very closely parallel to the optic axis of the laser emerges. The divergence angle is only determined by the properties of the laser beam, rather than by the boundaries of the system. It is usually in the milliradians range.

Beam divergence can be reduced by expanding and collimating the beam by a factor inversely proportional to the diameter of the expanded beam. It is normally measured by the half angle subtended by the diverging beam.

The radius of the laser beam is normally measured as half the distance between points at which the amplitude is 1/e that of the centre.

3.3. Energy and power output.

In lasers the intensity of the beam can be very great, since we have a concentration of emitted power almost in just one direction. Laser power is concentrated in a small diameter coherent beam.

The output of the laser may be continuous (CW=Continuous Wave) or pulsed. The pulsed output may vary from a single pulse to a series of repetitive pulses giving an almost continuous output. Since the duration of pulses can be very short (10^{-9} s), the power output can be very large (1 MW) even if the total energy is very small (10^{-3} J).

The magnitude of the power output is governed by many factors as:
a) laser transition,
b) method of excitation,
c) diameter and length of the laser,
d) rate of heat dissipation in laser host and pumping source and
e) overall gain of the cavity.

3.4. Mode structure.

The electromagnetic field generated inside the laser cavity is constrained to take up only certain allowed configurations or modes consistent with the boundary conditions. This boundary conditions (shape of mirrors, diameter of tube, inner aperture, Etc.) select the "modes" of vibrations, characterized by very defined intensity distributions within the beam. The classification of modes is based on waveguide modes at microwave frequencies, transverse electrostatic (TE) and transverse electromagnetic (TEM) modes. The designation for mode numbers on rectangular symmetry is normally used according to the convention TEM_{xy}, where the first subscript refers to the x axis and the second to the y axis. The mode number is always one higher than the number of illuminated zones. The zero-order mode TEM_{00} has an ideal Gaussian distribution. Mode selection is possible by varying the curvature of mirrors, by including appertures, resonant reflectors (reflection grating) or selective absorbers into the resonator.

4. TYPES OF LASERS.

Here we will only summarize the main characteristics of the most used lasers for applied laser tooling:
- Neodymium laser
- CO_2 laser
- Excimer laser

4.1. Neodymium laser.

Neodymium doped laser materials have a characteristic output wavelength at 1.06 μm. They are solid state lasers in which the host material is Yttrium Aluminium Garnet (YAG), other garnets and glass. It works according to a four-level scheme, either continuous or pulsed. YAG crystals are mainly used for materials processing because one can achieve higher powers. Continuous output of over 1000 W has been obtained from a single YAG crystals at efficiencies up to about 4% with divergence of about 3 milliradians (Ref.6). Fig.4 shows a typical elliptical configuration used as pumping scheme, in which the linear pumping source is located along one principal axis and the laser rod along the other.

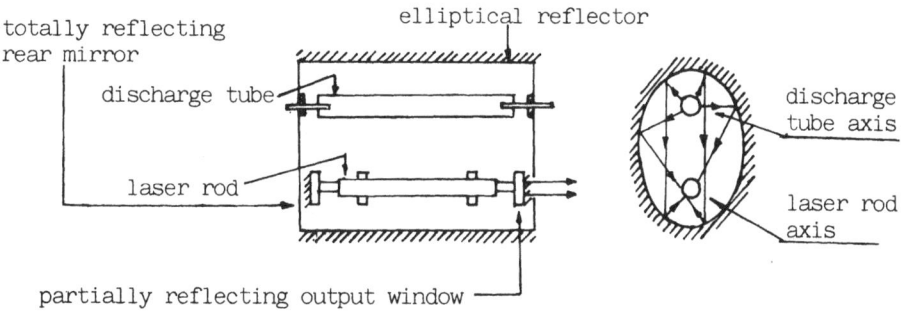

FIGURE 4. Neodymium laser with elliptical reflector.

4.2. CO_2 laser.

As most gas lasers, the CO_2 laser is excited by an electric

52

discharge inside the laser cavity. Different schemes are shown
in Fig. 5 used to obtaine a wide range of laser output power.
Here the electric discharge is influenced by the gas, gas
pressure, current and gas composition. It is a molecular laser
excited normally by a glow discharge. Cavity walls are normally water cooled borosilicate tubing. Fig. 6 shows the energy
transition scheme, with which very high efficiencies can be
obtained, typically of 18%. A large number of possible transitions exists over the band 9 - 11 µm, with a mean value around
10.6 µm. The output power is proportional to the length as
indicated in Fig. 5 for each type of gas flow configuration.
The laser gas is a mixture of 6% CO_2, 12% nitrogen and 82%
helium. This laser type is the most powerful one up to date.

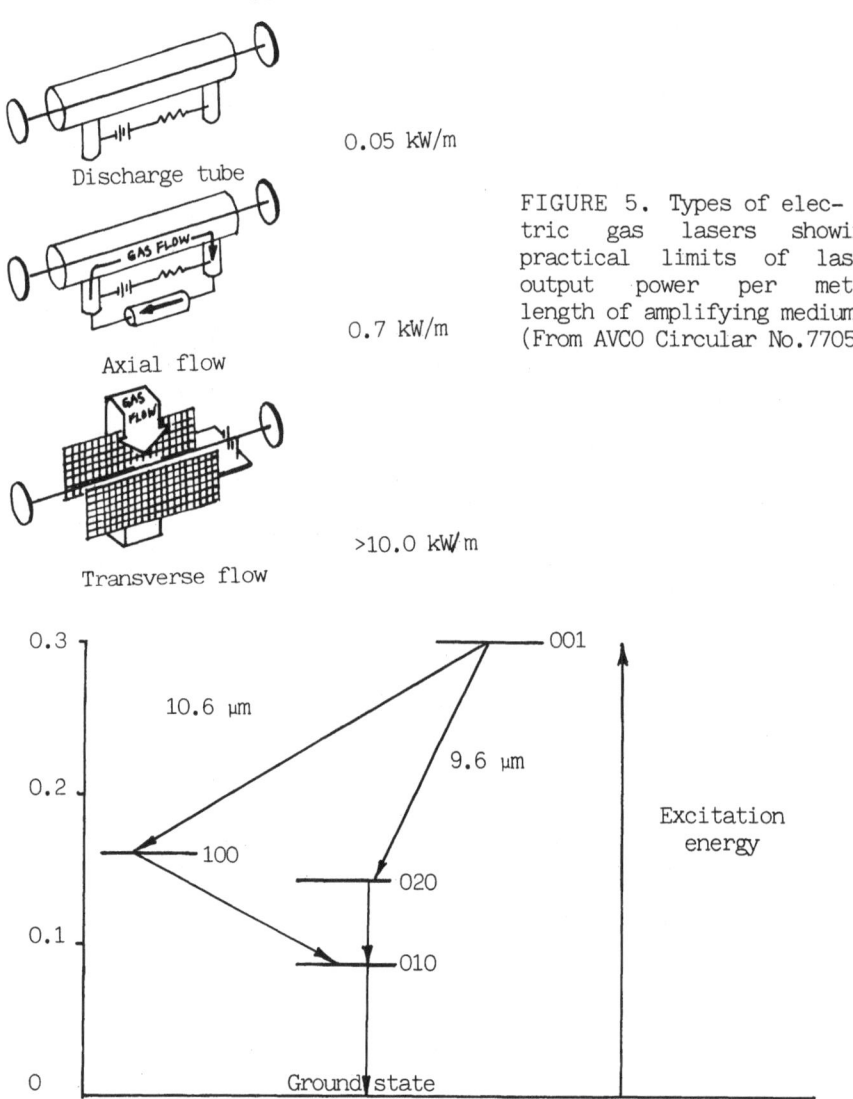

0.05 kW/m

Discharge tube

Axial flow

0.7 kW/m

FIGURE 5. Types of elec-
tric gas lasers showing
practical limits of laser
output power per meter
length of amplifying medium.
(From AVCO Circular No.7705)

>10.0 kW/m

Transverse flow

FIGURE 6. CO_2-laser level scheme.

4.3. Excimer laser.

It is a rare gas - halide laser, which is becoming the dominant source of laser light in the near and middle ultraviolet wavelengths. Excimers are molecules formed of an excited atom and a ground state atom. An example could be XeF with laser transitions at 3510 and 3530 Å. They can produce pulses containing kilo-joules of energy. Power levels are lower than for CO_2 lasers. See Ref. 7d for more details and further references.

REFERENCES

1. Maiman TH: Stimulated Optical Radiation in Ruby Masers, Nature 187, 493 (1960).
   Schawlow AL and Townes CH: Infrared and Optical Masers, Phys. Rev. 112, 1940 (1958).
2. Einstein A: On the Quantum Theory of Radiation, Phys. Z. 18, 121-128 (1917).
3. Ross D: Lasers, Light Amplifiers and Oscillators. Academic Press 1969.
4. Weber H and Herziger G: Laser, Grundlagen und Anwendungen. Physik Verlag 1972.
5. Other laser books:
   a) Lengyel BA: Laser Light. Wiley Interscience 1971.
   b) Yariv A: Quantum Electronics. John Wiley 1975.
   c) Verdeyen JT: Laser Electronics. Prentice Hall 1981.
   d) Harry JE: Industrial Lasers and Applications. McGraw-Hill 1974.
   e) Svelto O: Principles of Lasers. Plenum Press 1976.
6. Koechner W, DeBenedictis LC, Matovich F and Mevers GE: Characteristic & Performance of High-Power CW Krypton Arc Lamp for Nd:YAG Laser Pumping, IEEE J. Quantum Electron. 8, 310-316 (1972).
7. References about special laser types:
   a) Patel CKN: High Power Carbon Dioxide Laser, Scientific American 219, 22-23 (1968).
   b) Cheo PK: CO_2 Lasers: Lasers, Vol.3, Levine AK and DeMaria A (eds), Marcel Dekker, NY (1971).
   c) Duley WW: CO_2 Lasers: Effects and Applications. Academic Press, New York 1976.
   d) Hecht J: Excimer Laser Update, Laser & Applications 2, No. 12, 43 (1983).
   e) Klauminzer G: Cost Considerations for Industrial Excimer Lasers, Laser Focus 21, Nº 12, 108 (1983).
   f) Danielmey HG: Progress in YAG Lasers: Lasers, Vol.4, Levine AK & DeMaria A (eds), Marcel Dekker, NY 1976.
8. Klauminzer GK: Twenty Years of Commercial Lasers, Laser Focus 20, Nº 12, 54 (1984).
9. Kogelnik H & Li T: Laser Beams and Resonators, Appl. Opt. 5 (10), 1550-1567 (1966).

PROCESS AND PHYSICAL ASPECTS OF CONTINUOUS WAVE LASER PROCESSING

R.C. CRAFER and P.J. OAKLEY
The Welding Institute, Abington, Cambridge, UK

INTRODUCTION

Laser processing embraces six major activities, namely 1) generation of light within the laser, 2) transmission to the workpiece, 3) quality assurance of laser beam parameters as delivered, 4) absorption at the workpiece, 5) the process itself, and finally 6) quality assurance of the finished product. Although each activity may be considered in isolation, one should ideally consider all six as parts of an overall laser process chain. The reasoning behind this integration of activities is that the activities are interrelated. As an example of what could occur to upset the normal state of affairs, consider a local change in the chemical composition of the material being processed giving rise to a high electron density at the absorption interface and plasma frequency reflections. The excess reflected light reaches the laser and changes its mode structure which affects both beam measurements and further absorption. In practice unwanted feedback can be minimised by careful design and consideration for each activity; however certain feedback is highly necessary, such as quality assurance between activities 3 and 1 and 6 and 1. With this in mind, this chapter examines each activity from a process engineering viewpoint with particular regard to eliminating or optimising feedback as appropriate, and takes its examples from continuous power welding and surfacing using continuous carbon dioxide lasers.

GENERATION

The high intensity light beams used for materials processing are generated by devices called LASERS. The letters stand for LIGHT AMPLIFICATION by STIMULATED EMISSION of RADIATION, a process suggested by Einstein and first realised practically by Maiman. The physics of the laser generator has already been outlined in Dr. Leon Fong's chapter of these proceedings and will not be elaborated here. From a process point of view the LASER can be considered as an energy transformer, taking in high power low grade energy from the 'pump' source and putting out lower power higher grade energy as a collimated, coherent laser beam. As with all transformers, some energy is lost in the transformation and manifests itself as heat in the LASER device. It is this heat which effectively limits the performance of industrial lasers.

In the case of the carbon dioxide laser, the low grade input energy is provided by an electrical power supply, the transformation occurs in a low pressure mixture of carbon dioxide, nitrogen and helium

Soares, O.D.D., Perez-Amor, M. (eds), Applied Laser Tooling. ISBN-13: 978-94-010-8096-5
© *1987. Martinus Nijhoff Publishers, Dordrecht.*

gases, and a pair of mirrors fashions the stimulated output into a
collimated monochromatic light beam having an infrared wavelength of
10.6 microns. Figures 1a-c illustrate commercial carbon dioxide laser
devices of 1200W to 10000W.

TRANSMISSION

The purpose of the transmission is to relay the high power beam from
the laser generator to the workpiece in a manner appropriate to the
application. With the highly developed state of modern day laser
generators, the transmission is perhaps the most critical aspect of
the laser processing chain, and applications can succeed or fail
depending on the suitability and efficiency of the transmission. A
proper study of transmissions begins with the laser generator itself.
There are several methods by which the beam power is coupled out of
the laser generator and launched towards the workpiece. All known
methods can distort and perturb an initially coherent and collimated
beam, thus reducing its potential effectiveness for the application.

At low and medium powers, up to say five kilowatts, the laser is
likely to be fitted with a 'stable' resonator, from which the beam is
extracted through a partially reflecting transparent plane window.
For technical reasons related to the long infrared wavelength of
carbon dioxide lasers, the window material could be either germanium,
gallium arsenide, cadmium telluride or zinc selenide, all of which
transmit radiation in the 10.6 micron region of the electromagnetic
spectrum at which these lasers operate. Being semiconductors with
narrow energy gaps between valence and conduction bands, they all
exhibit strong absorption of light when heated, resulting in thermal
lensing or, in extreme cases, structural failure. The major
differences between the materials are the temperatures at which
absorption occurs. At one extreme, germanium begins to absorb at less
than 50 degrees centrigrade, while zinc selenide remains essentially
transparent up to several hundred degrees. The actual temperature
obtained is a balance between heat absorption and losses. Absorption
occurs via impurities within both substrate material and
reflective/anti-reflective coatings, and also by electronic absorption
from conduction band electrons. Cooling occurs predominantly via
convection at the window faces and conduction into a cooling medium at
the edges. In practice, by judicious choice of material and coatings
for the size and power of laser, beam distortion can be minimised,
certainly at powers up to five kilowatts, but long term effects such
as surface contamination can cause unexpected absorption, and such
possibilities should always be anticipated in maintenance schedules.

At higher powers, of order ten kilowatts and above, it is common
practice to use the so-called 'unstable' resonator. Here, coupling is
achieved via mirrors which being metallic and of high thermal
conductivity can handle much higher powers than semiconductor windows.
However it is still necessary to transmit the beam from the generator
environment to the working environment, and this still requires some
form of window. Two types are in common use, solid alkali halide
windows such as potassium chloride, which are relatively inexpensive
and are sufficiently large to fit the 'unstable' beams, and
aerodynamic windows, where the beam is brought to a preliminary focus

through a series of differentially pumped apertures. In the former case, lensing can occur though not in the catastrophic manner of semiconductors, while aerodynamic windows can refract the beam through density gradients at the pumped apertures.

Having extracted the beam from the laser, it must be directed to the workpiece. This is almost exclusively accomplished by means of mirrors.

There are four basic variations as follows:-

1. Stationary laser – stationary mirror(s) – moving workpiece
   Suitable for small, low inertia workpieces, or for large and immobile lasers.
   Advantages – simplicity of construction and constancy of beam path – allowing the designer to miminise transmission distortion.

2. Moving (lasers and mirrors) – stationary workpiece
   Suitable for low power low inertia laser with large workpieces
   Advantages – simplicity of construction and constancy of beam path – allowing the designer to minimise transmission distortion. Optically similar to 1.

3. Stationary laser – moving mirrors – stationary workpiece
   The 'ideal' solution which places no limitations on laser generator or workpiece but which poses difficult problems for the designer of the optical-mechanical system. Many designs have been attempted all with their own advantages and problems. Perhaps the most exciting recent development has been the articulated beam tube which can be driven by a robot arm. This is the nearest approach so far to a flexible cable capable of carrying a multikilowatt beam from generator to workpiece.

4. Stationary laser – moving mirrors – moving workpiece
   A solution which often achieves the objectives of 3 with less complication, provided the workpiece is moveable.

Once in the vicinity of the workpiece, the beam has to be fashioned into a form suitable for the application. For welding and cutting at low powers, lenses are commonly used to bring the beam to a focus. Semiconductor are the preferred materials here because the high refractive indices permit small surface curvatures and low spherical abberations, giving rise to small focal spots and high intensities at the workpiece. However, semiconductor lenses are expensive, and in less critical applications or where lower intensities can be tolerated, alkali halide lenses can be employed, although anti-reflection coatings with long term stability have yet to be demonstrated. For high power welding and cutting applications mirrors are usually employed as focusing elements. For technical reasons, off-axis operation is commonly employed which sometimes requires specialised aspherical designs, such as the off-axis paraboloid.

For heat treatment applications there are three types of solution namely defocusing, beam distribution and raster scanning. Beam redistribution involves juxtaposing different parts of the unfocused beam to form an intensity profile which provides an approximately uniform treatment depth when moved across the workpiece. Special plane mirrors or aspheric components such as axicons are used to accomplish this. Raster scanning involves moving (vibrating or rotating) one or more of the transmission mirrors so that a conventionally focused spot traces out a raster pattern of the required time-averaged intensity. Many useful transmission systems have been constructed but there are also many problems to be avoided. Some of them are listed in the following paragraph.

Every optical element in the transmission will reduce the quality of the beam as delivered to the workpiece. All mirrors will both absorb light and scatter it in unwanted directions. The loss per mirror may only be a few percent, but in a multimirror system such losses quickly accumulate. With the exception of perfectly figured aspherical mirrors, all other curved mirrors are subject to aberrations. Lenses, in addition to these problems, also reflect an unwanted amount of light from each surface, although this may be reduced to acceptable levels on some materials by anti-reflective coatings. In addition to the above, if any of the transmission components become heated by the beam, then their shape will alter and the beam quality will be further reduced. Finally, in some high power applications, the atmosphere between the optical components may itself become heated and thus reduce further the quality of the beam.

BEAM QUALITY ASSURANCE

The beam emitted by a continuous power laser can be characterised by three parameters: power, diameter and intensity distribution. Beam power is measured routinely in high power lasers used for materials processing. It is a principal variable for processes such as welding, cutting, surfacing, etc. Beam diameter is defined by the the optical resonator; however, measurement of a partially or fully focused beam is also of interest. The intensity distribution within the beam is a function both of the resonator and the relative alignment of the optical components. The distribution is usually axisymmetric and can be measured by the beam profile along a radius or diameter. Given the beam profile, a diameter may be deduced either in terms of the fraction of beam power passing within it, or in terms of points where the intensity is reduced to a given fraction of its peak value. Although making fine adjustments of the laser optics by reference to intensity distribution as well as laser power is desirable, this is not easily achieved with present measurement techniques. As lasers become established as production machines for a range of applications, the monitoring of these parameters will become increasingly important for process and quality control.

Beam parameters can be measured by using the total beam, or by sampling a known fraction of the beam. Total beam measurement is the simplest technique but has the disadvantage that measurements are not available during processing. Three methods of sampling part of the beam, usually a few percent, are shown in Fig. 2. As only a small sample of the beam is monitored, techniques with lower power capabilities can be used. All three provide the advantages of monitoring the beam during processing, but can reduce the accuracy of measurement because of uncertainties in the fraction of the beam sampled. Each method has additional disadvantages. A static beam splitter (Fig. 2a) is a delicate optical component and, as it is typically manufactured from dielectric coated zinc selenide, it can be expensive. It will absorb some power from the transmitted beam, and may not be suitable for use at higher power densities. The chopper wheel (Fig. 2b), which consists typically of a rotating 45 degree gold plated copper blade, cuts across the beam and hence there is a modulation of the laser beam which may be significant when processing material at high speeds. The use of a coated dielectric mirror, or a small aperture in a conventional mirror at the non-output end of a laser resonator, will also sample a fraction of the beam (Fig. 2c). A coated dielectric mirror is again an expensive and delicate component and a similar restriction on maximum allowable power density may be encountered. An aperture in a conventional fully reflecting mirror necessitates a known intensity profile within the laser cavity if an accurate fraction is to be sampled. This third technique cannot, of course, be applied to lasers operating in an oscillator/amplifier configuration.

Total beam power is usually measured by calorimetric techniques using either air or water cooled calorimeters. Air-cooled devices are restricted to lower powers than those cooled by water. Figure 3 shows a water-cooled calorimeter, where the beam is absorbed on an internal conical surface. Lower powers (<1kW) can be measured on water-cooled calorimeters with a flat absorbing surface. A laser system is usually purchased with a calorimeter designed for the beam size and output power; however, free-standing calorimeters are commercially available for powers up to 1kW. Measurements can be made either by directing the total beam into a calorimeter, or by sampling a known fraction of the beam. When measuring the total beam a mirror is used to switch the beam from the workstation to the calorimeter. The calorimeter therefore provides a beam dump while the laser is operating but not processing. A combination of in-process measurement and total beam calorimetry provides a comprehensive power measurement system for carbon dioxide lasers.

Diameter and Intensity Distribution

Heating or charring sensitive materials is the simplest method of looking at the profile of an unfocused or partially focused beam. Materials used for this include wood, firebrick and asbestos compounds. This technique gives an indication of beam diameter and symmetry. It is, however, a technique where the affected area, and hence the measured beam diameter depends on both power and exposure. It is therefore more suitable for comparative measurements. Courtney and Steen describe the reservations about this technique in greater detail (1). Incandescent mode plates and materials that are chemically sensitive to the laser wavelength can be used with lower power lasers or small fractions of high power beams but are subject to similar reservations.

Measurements made by vaporising acrylic materials are also subject to reservation; however, more information can be obtained than with

heating or charring because of the possibility of vaporising a deep
region in the acrylic and hence making a three dimensional print,
although the technique is again more comparative than absolute. A
deep acrylic print of an unfocused beam is shown in Fig. 6. This type
of print gives a better measurement of beam diameter and intensity
than a shallow print. An alternative technique of using expanded
polystyrene for making deep burn prints is described by Meyerhofer
(2), together with calculations for analysis of the results. Because
of the speed at which this material vaporises, the technique is
restricted to short exposure times using low power lasers of the order
of 100W or less.

Courtney and Steen have also overcome the dependence of measured
beam diameter on laser power and time by measuring the time for the
temperature of a laser heated spot to rise to 90% of its equilibrium
value. This time is compared with predictions for a Gaussian laser
beam, and an effective Gaussian beam diameter deduced.

Pyroelectric detectors can be used to obtain intensity profiles. As
pyroelectric devices respond only to variations in power, it is
necessary either to move the device through the beam or use a chopper
wheel. The devices are fragile and susceptible to thermal overload
and are therefore limited to lower power densities - typically 1W/mm .
Various sophisticated systems have been reported combining pyroelectric detectors with beam scanning optics (3,4). Sepold has
described the use of a Nipkow disc technique (5) with a 5kW CO_2 laser.
Pyroelectric vidicons have been used for real-time measurement of beam
intensity profiles. These devices are television tubes with
pyroelectric cathodes. The measurement of low power beams has been
described; however, improvements in resolution and sensitivity are
required and their use has not yet been reported for high power cw
lasers.

Beam diameters can also be measured by their effect on a workpiece.
Partially focused beams are used to effect solid phase transformations
or melting in laser surfacing processes. Where melting has occurred
the treated width can be measured directly on the surface. However, a
more accurate measurement of a molten region and a measurement of a
solid phase transformed region can be made on a polished and etched
cross-section.

Shallow tracks made by traversing an angled acrylic sheet across the
laser beam will give the position of focus, but because of the high
power density no information about the beam diameters can be deduced
close to focus.

Of particular interest in welding and cutting are the focal diameter
and profile of the beam. Two methods are described in (6) and (7).
The devices involved are complex and the results require interpretation and interested readers are referred to the references.

A typical output signal from one of these high intensity beam
scanners is shown in Fig. 5. This signal is derived from a three
dimensional beam slice (Fig. 6) and must be used to reconstruct the
two dimensional intensity profile using a mathematical process known
as deconvolution. A similar problem arises when monitoring electron

beams and a technique based on a method due to Harker has been used at
The Welding Institute (8). This method is tolerant to noise and
signal inaccuracies but requires a narrow sample width (typically 20:1
beam diameter to sample width ratio) to perform accurate intensity
profile reconstruction. In the laser case it is more difficult to use
such a high ratio because the low signal voltage levels are obscured
by noise, thus ruling out Harker's method. An alternative, more
limited method of interpretation is sometimes used in which
'reference' results are compared with actual displays.

ABSORPTION

The first stage of any material absorption mechanism is when the
beam strikes the material surface. In the case of an insulator, the
crystal lattice composed of positive and negative ions becomes
strained by the electric field component of the incoming light beam.
Upon reversal of the field the strain energy is released and helps the
beam to propagate further. Depending on the electrical 'stiffness' of
the insulator, some energy propagates forwards into the material and
some is reflected backwards. This is the classical behaviour of
insulators such as potassium chloride and cold semiconductors such as
germanium and zinc selenide. The release of energy back into the
transmitted or reflected beams implies no losses, and accounts for the
excellent window behaviour of these materials. In the case of a
metal, which can be regarded as a crystal lattice filled with free
electrons, the electrons gain energy from the electric field, and
provided no collisions take place within one cycle of the alternating
field, no energy is absorbed. If, however, an electron collides with
a lattice impurity, then its energy is used to heat up the crystal
lattice and the beam energy is diminished. In practice, most metals
behave in this latter fashion for light wavelengths exceeding about
0.5 microns, which includes both YAG and carbon dioxide lasers. A
full theoretical analysis shows that most of this absorption occurs in
a narrow surface layer, typically 30 nanometers thick, or a few atomic
layers, so that laser absorption in metals is truly a surface effect,
and that the absorbed fraction may vary from about one percent for
copper to tens of percent for steels. A variation of this absorption
process is used for solid phase surface treatment or surface alloying.
Clearly even a few tens of percent absorption is insufficient and
inefficient, so a matt surface coating is applied, e.g. colloidal
graphite. Light not absorbed by any one particle is likely to be
reflected onto other particles and so on, thus greatly enhancing the
absorption process. The heat absorbed by the coating is then
transferred to the metal surface by conventional methods.

For welding, a second stage of absorption is required as coatings
cannot be used since they would be incorporated into the weld metal
with adverse characteristics. In this case it is a prime requirement
that the first stage of absorption is able to bring the metal surface
to its boiling point. Low absorption metals such as gold and copper
will clearly be more difficult to vaporise than high absorption
steels, and indeed only the highest power lasers are capable of
welding these materials. Once vaporised, some of the metal electrons
become free, a process called ionisation. These free electrons can
absorb energy directly from the incoming light beam by a process known
as Inverse Bremsstrahlung, resulting in higher temperatures, more

Ionisation and increased absorption. The rapidly increased absorption
vaporises the surface underneath, which recedes and produces a cavity
through the depth of the metal. This cavity is known as a keyhole and
forms the basis for deep penetration laser keyhole welding (Fig. 7).
The keyhole is a liquid-lined cavity with a very hot ionised gas core.
Surface tension and gas pressure holds the liquid in place against
gravity and other forces, while light absorption in the central gas
core keeps the process going. Welding takes place by moving the beam,
and hence gas core through the metal and melting the adjacent
material. In simple terms the fluid flow of molten metal past the
keyhole can be ascribed to an imbalance of surface tension forces, but
for a more detailed analysis the reader is referred to references 9
and 10. In practice, keyhole absorption is very efficient and
accounts for the majority of the beam power. In extreme cases, the
hot metal vapours escaping from the keyhole can actually absorb the
beam above the metal surface, causing keyhole collapse and poor
welding performance. This is known as the plasma effect, and various
methods have been used to overcome it, principally the use of
auxiliary gas jets to suppress and remove the absorbing plasma cloud.

LASER WELDING

Carbon dioxide lasers between 1kW and 20kW are becoming increasingly
accepted as production tools for welding. The highest power machines
are capable of welding, in one pass, thicknesses of at least 25mm of
steel whilst at the other end of the scale metals of foil thicknesses
can be welded using low power lasers. This makes the process suitable
for a wide range of applications, particularly where numbers of
similar assemblies are to be welded, allowing full automation of the
process and very high production rates to be achieved. Actual and
potential applications range from electronics to heavy fabrication,
although at present the majority of applications require weld
penetration depths of 3mm or less. Applications are also restricted
by the materials, with the higher reflectivity metals, such as
aluminium alloys, copper and gold, being difficult or impossible to
weld with a carbon dioxide laser.

Process Parameters

Five process variables can be identified when using a particular
continuous wave (cw) laser for welding. These are power, traverse
speed, diameter of the focused beam, depth of focus, and type of gas
shielding. The two beam parameters are interrelated, both being
dependent on the focusing system used. The type of laser used
dictates other variables such as the wavelength of the beam and
intensity distribution. Sophistications of the process, such as the
use of wire feed or beam oscillation, introduce further variables
which require quantifying.

Measurement of the beam parameters (power, diameter, intensity
distribution) have been described earlier in this chapter. For a
laser to weld efficiently it is necessary to operate with the beam
focused at or close to the workpiece surface. The diameter of the
focused beam and the depth of focus quantify the beam in the focused
region and, whether based on lenses or mirrors, these parameters are
related to the f. number of the focusing system. For any particular

beam diameter, this means that these parameters are related to the focal length of the focusing system. Although researchers have investigated the effects of different focal lengths on weld penetration and quality for particular lasers, there is insufficient data available to formulate generalised rules (11). It is, however, possible to make some comments. A focusing system with a short focal length will result in a smaller diameter of focused beam than a system with a long focal length. However, the penalty for using a short focal length is a much reduced depth of focus. Short focal lengths (<150mm with a 30mm diameter beam) are therefore more suited to the welding of thin materials (<4mm) where high power density is required to maintain the weld keyhole at high speed. The greater depth of focus of a longer focal length system is more efficient for welding thicker materials at comparatively slow speeds. Practical considerations of access with very short or very long focal length systems, and the vulnerability of short focal length systems to damage by reflections or spatter from the weld have also to be taken into account when selecting the focusing arrangement. In the latter respect mirrors are usually better than lenses.

The gas shield has several functions. It gives some protection to optical components from   fume and spatter, ensures effective transmission of the beam through the hot region of gas just above the workpiece, and protects the molten material from reaction with the atmosphere. It is generally agreed that helium is the best gas to use, but it is expensive, so there is much work currently in hand to investigate other gas mixtures (12,13).

The level of gas shielding required is dependent on the specific welding application, but in general the slower the welding speed the more opportunity there is for the atmosphere to react with the weld metal, so a more comprehensive gas shield is required. The gas shield must also provide plasma control, particularly at high powers and low welding speeds. At low welding speeds a plasma of metal vapour escaping from the top of the keyhole interacts with the laser beam to lessen the penetration capability of the beam. A high velocity jet of helium is usually directed at the top of the weld to blow away this plasma (14). Figure 8 shows in schematic form a plasma control device. Another technique that has been proposed is to pulse the laser beam at a high frequency so that the duration of each pulse is shorter than the time taken to generate the plasma, thus enabling the beam to penetrate the material efficiently (15).

Seam Tracking and Fit Up

If the components to be welded are made to accurate tolerances, then the work handling used must be capable of good positional acuracy to ensure accurate alignment of the joint line with the laser beam. For example, the fusion zone width of a butt joint in 2mm thick low carbon steel is less than 1mm when welded at a speed of 3m/min and a laser power of 3kW. In this instance, if the laser beam focus spot is positioned 0.5mm to one side of the joint line, the fusion zone would miss the joint. The actual tolerance to joint alignment depends on the particular application, but is invariably less than 1mm and in general the positional accuracy of the work handling equipment must be smaller than the minimum fusion zone width produced by the welding conditions employed.

On the other hand, in large volume production the manufacturing tolerance of component parts cannot always match that required for consistent positioning of the joint line, so that despite the use of accurate work handling a missed joint will sometimes occur. Under these circumstances seam tracking is necessary to adjust the position of the joint with respect to the laser beam. No current work on seam tracking for laser welding is known but The Welding Institute is currently developing a seam tracking system for mechanised MIG and TIG welding and this may also prove suitable for laser welding (16).

Because the laser beam focused spot size is so small, close fitting butt joints are necessary if consistent quality autogenous welds are to be made. A gap of, say, 3mm would pose little difficulty to most arc welding processes, but a laser beam would pass straight through it. The size of gap that can be tolerated depends on the thickness of the material and also whether or not undercutting of the weld bead can be tolerated. As an example of the importance of fit-up, good quality butt welds without undercut cannot be made in 2mm low carbon steel at high speeds by conventional laser welding if there is a gap much greater than 0.1mm between the abutting faces. In thicker materials (i.e. >6mm) larger gaps can be tolerated, but gaps of more than 0.5mm are unacceptable in all cases.

Although it is an added complication, the introduction of a filler wire from an automatic wire feed system increases the tolerance to gap considerably. At The Welding Institute single pass laser welded butt joints have been made in 8mm thick structural steel with joint gaps of 2mm. Multipass welds with even larger gaps have been made in 25mm thick plates. The feasibility of using a filler wire for welding low carbon steel materials down to 2mm thick has also been established. Another advantage of using a filler wire is that the chemistry of the weld metal can be changed to control the properties of the weld to suit subsequent processing and application requirements (e.g. toughness).

LASER SURFACING

Lasers are inefficient bulk heating devices and their use for through thickness heat treatment has not proved feasible. However they are effective at heating discrete areas very rapidly, and are therefore suitable for certain heat treatment and surfacing applications. The laser can produce a high power density, matched only by electron beam equipment. When using a conventional heat source, such as an oxy- acetylene flame for surface treatments, the relatively low power density allows heat to be conducted into the bulk of the component while the surface is reaching the required temperature, thus heating the surface inefficiently. The high power density laser beam heats the surface much more rapidly, reducing time for heat conduction into the bulk of the component. The laser is a localised controllable heat source, and can therefore be utilised to treat a component efficiently by heating only discrete areas.

The processes with which laser heat treatment and surfacing techniques must compete include a wide range of comparatively low cost

conventional processes. The use of laser techniques must therefore
offer significant advantages.

The general advantages are:

- treatment can be localised to the required area
- heat input is relatively low, giving minimum thermal distortion of
  the component
- no, or very little machining required after treatment
- complex component shapes can be treated
- the laser beam can be directed, by mirrors, to treat inaccessible
  areas of components
- most treatments are rapid and can be incorporated into a
  production line

Specific advantages can be found with each of the different laser
surfacing processes, and secondary, sometimes unexpected, benefits
have been found when these processes are considered for a particular
application.

Process Classification

The methods of laser surface treatment of materials have been
classified into three types, namely those involving heating, melting
or shocking, Fig. 9, (17). The processes can also be divided into
those relying on a metallurgical change in the surface of the bulk
material - transformation hardening, annealing, grain refining,
glazing and shock hardening - and those involving a modification to
the chemical composition of the surface by addition of new material -
alloying and cladding. Subsequent to these classifications an
additional technique called particle injection has become the subject
of research. This involves the addition of solid powder into the
molten surface of the substrate, and hence can be classified with
alloying and cladding.

Unlike laser welding and cutting, where the beam diameter is focused
to a minimum size at the workpiece surface in order to give a narrow
fusion and heat affected zone, most laser surfacing processes require
a reduction in power density and preferably treat a wider band of the
component. A range of beam handling techniques from defocusing to
rasters and multi-faceted mirrors can be used. The use of a defocused
beam is the simplest way to reduce power density and give the required
treatment width. However, it is difficult to obtain a uniform power
density across the beam. A defocused gaussian beam, for example, will
have a maximum power density at its centre and thus will produce a
maximum treatment depth at the centre of its track across a component.
A more complex beam handling system can be utilised to give a uniform
power density, but this usually is reflected in higher cost optical
components, together with less flexibility and reliability.

Metal surfaces are good reflectors, particularly at the 10.6 micron
wavelength of carbon dioxide lasers. This high reflectivity, and
hence poor coupling of the energy to the workpiece, can limit the
effectiveness of laser metal working processes. The absorption of
infrared radiation does, however, increase with increasing surface
temperature and shows a sharp increase once the material is molten.

This reflectivity is overcome when laser welding and cutting by the beam being absorbed in a plasma filled 'keyhole'. During surfacing processes, where the component surface is as undisturbed as possible and its temperature is lower, it may be necessary to treat the component surface to increase its absorption of the laser beam, particularly for processes that do not involve surface melting. To be effective, the coating applied to the component must highly absorb the laser wavelength and must be very thin to allow effective transfer of energy to the component. Typically used coating materials are flat black paint, colloidal graphite, or zinc and manganese phosphates. Powdered materials on a component surface will facilitate the absorption of the laser beam and therefore if material is added as a powder in alloying or cladding the application of a coating to enhance absorption may not be required.

Transformation Hardening

Transformation hardening is achieved by moving the beam across the surface of, for example, a carbon steel component. A thin surface layer of material is rapidly heated to above the austenitising temperature in the short time that the beam is incident. Once the beam has moved on, this surface layer is quenched by conduction of heat into the still cool bulk of the component, producing a hard thin martensitic layer. The process is therefore self-quenching and an external liquid quench is not usually required, although highly contoured thin components may need quenching or processing on a chill block. Typical hardened depths are 0.5-1.0mm and the width of the hardened track depends on the incident beam at the material surface, but typically may be between 2 and 20mm. Power density and inter-action time must be carefully controlled in order that the required treatment depth of the material reaches the austenitising temperature but the surface does not melt, as this would produce an uneven surface and necessitate subsequent machining.

The range of alloys that can be transformation hardened by laser techniques covers those that can be hardened by more conventional methods. As expected, the response of a steel to hardening increases with increasing carbon content, and in addition, because of the high cooling rates, comparatively low carbon steels (~0.2%) will show some hardening. This compares to a practical lower limit of approximately 0.3%C for more conventional hardening techniques. Some materials do, however, present difficulties because of the short diffusion times. A significantly longer interaction time may be required to harden a cast iron or steel with a coarse structure and wide pearlite spacing than is required to harden a material with the same composition but a finer microstructure. For materials in which the solid state diffusion is not sufficient to obtain complete solution of carbides or carbon, laser transformation hardening can be achieved by a melting and solidification process. This has the disadvantage of producing a roughened surface that will probably necessitate a subsequent machining operation, and also increase the risk of cracking, particularly with cast irons. Usually a transformation hardened region is introduced on a component to combat wear, but it can also increase fatigue life as the martensitic transformation induces surface compressive stresses.

Tempering and Annealing

Controlled heating and cooling can temper or anneal a material just as it can harden it. If the quenching rate is too slow when transformation hardening irons and steels, then a transformation product softer than martensite will form and maximum hardness will not be obtained. Similarly, by controlling the power density and interaction time of a laser beam with certain non-ferrous metals, it is possible to anneal discrete areas on the surface. However, the relatively short interaction times of laser surfacing limit the ease of application of these processes and no production usage is reported.

Alloying and Cladding

Laser alloying and cladding both rely on the melting of the component surface and the introduction of additional material. This material can be in powder or solid form (i.e. rod, wire or sheet) and can be preplaced or fed in as the process progresses. With alloying the objective is to get complete dilution of the added material in the molten surface layer of the substrate, whereas for cladding the minimum dilution of the added material by the substrate is aimed for. For both processes the depth of the surfaced layer is usually 0.5-2.0mm. However, with cladding there is the possibility of making several passes and thus building-up thicker layers.

The laser alloying process is still in the research and development stage and, although some applications have been assessed, there are no reports of its use in production. Much of the reported research has involved the surface alloying of low carbon steel with a range of elements (for example Cr, C, Ni, W, V, Mn) but there is also interest in increasing the silicon level in aluminium-silicon alloys (18). Although steel and aluminium alloys are the materials most often considered for process development, there are no process limits, other than the inefficiency introduced by high reflectivities, to the substrate materials or alloying elements that can be used. Applications that have been assessed for laser alloying include components such as valves, valve seats and piston rings where again the specific nature of the laser process is exploited.

In adding a completely new alloy to the surface of a component, laser cladding must compete with techniques such as TIG, oxy-gas welding, and thermal spraying. The laser parameters must be controlled to minimise the melting of the base material, yet achieve a good bond between coating and component. A metallurgical bond will require some dilution at the interface but this can be controlled to give a total cladding dilution as low as 1-2% and typically 5-8% (18). This is lower than dilutions found with conventional fusion cladding processes, thus reducing the degradation of the properties of the clad layer. In addition the localised heating possible with the laser allows a minimum of cladding material to be applied, hence minimising post-treatment machining. Alloys used for cladding are usually cobalt, nickel or iron base, although, as with alloying, a wide range of materials could be used. The consolidation of a coating previously applied to the material surface by a thermal spraying method is possible, in addition to the more usual ways of adding material. Cladding is extensively used for applications involving metal to metal wear, impact, corrosion, erosion and abrasion. Laser cladding can also provide coatings to combat these types of wear, and with the versatility inherent in laser techniques can be used to treat very specific areas.

Powder Injection

A development of laser cladding and alloying is power injection. This technique involves melting a surface layer of substrate material and injecting particles into the molten material to improve surface wear properties. High melting point materials are chosen for the powder so that the particles remain in the solid phase and are incorporated into the molten surface layer as discrete regions. Research into this process is reported by the US navy (19) and their experiments involve the injection of tungsten and titanium carbide powder into stainless steel, and aluminium, titanium and nickel alloy substrates. There are no reported applications of the process yet.

Glazing

Laser glazing is the term given to the production of unique metallurgical structures by rapid quenching of molten alloys. Although metallic glasses can be produced with certain alloy compositions, the process usually results in a much reduced grain size, extended solid solubility and the formation of new phases. These structures may ofer high corrosion resistance, and an otherwise unobtainable combination of hardness and ductility. In order to achieve the very high cooling rates required, it is necessary to scan the focused laser beam over the substrate material surface very rapidly. This results in glazed tracks that are less than 1mm in width and depth, and hence surface coverage rates can be slow. The response of a range of alloys, particularly those applicable to aero-engine manufacture, has been investigated (20), and the researchers have developed a technique for building up bulk rapidly quenched material on a cooled mandrel using the laser. The production of rapidly quenched material by laser is however at a research stage. The effects of the unique microstructures on the properties of a wide range of engineering materials must be assessed and, as these microstructures are thermodynamically unstable, service temperatures and lives may be restricted.

Grain Refining

Surface melting by laser, without material additions or rapid quenching can be used to refine the surface structure of a component. The process is of particular relevance to cast alloys, where slow cooling rates in large castings can result in a large grain size with pores, and oxide and sulphide inclusions. Localised melting of the surface can give a much finer structure with a more uniform inclusion distribution. This has the potential of improved fatigue strength, corrosion resistance and wear characteristics of cast components. Although there is a wide range of possible applications, particularly those using high quality cast components, e.g. aero-engine, no practical use of the process has been reported.

PRODUCT QUALITY ASSURANCE

General Comments

Quality assurance of laser treated components is an important part of the process chain, particularly as automated techniques are commonly employed. The assurance techniques for welding and surfacing are similar in many respects, and this discussion is therefore limited to laser welding, although many of the techniques are also applicable to laser surfaced components.

Tests used to establish the quality of laser welds fall into two distinct categories depending on the reason for making the tests. Firstly, there are tests carried out to establish the feasibility of laser welding for any particular application, i.e. laboratory trials or acceptable trials to determine whether or not laser welding can produce welds meeting the requirements of the particular application. Clearly the tests used will depend on the application and could range from a simple visual examination to check weld integrity in a non critical component, to a full evaluation of defect levels and mechanical properties in a highly critical component in, for example, an aero-engine. Secondly, there are the tests used during production to ensure that welds of the required quality are being made. Laser welding is a high productivity process so tests used must be quick and simple. Both the above categories of test can use destructive or non-destructive testing techniques as appropriate but, of course, destructive tests can only be applied to a small percentage of production welds. Consequently, in production, control of machine parameters is very important. It should perhaps be emphasised that welding is usually the last of a series of operations involved in fabrication and the weld quality that is obtainable is a reflection of earlier processes which, for example, may produce variable fit-up and so on.

Currently in most applications the control of weld quality is limited to simple systems monitoring one or two process parameters (i.e. power and speed) and relying on good component preparation. Some simple batch testing may also be used. In future, however, the control of the laser manufacturing system should become more sophisticated, for example, altering welding conditions as workpiece fit-up varies and using automatic NDT techniques to accept or reject the welded components.

Non-destructive Testing Techniques

All of the commonly used non-destructive testing (NDT) techniques can be applied to laser welds, and the principal defects that require detection (lack of fusion, cracking, porosity) are the same as in arc welding. It is important to recognise that NDT cannot itself tell if a weld is acceptable; all it can do is detect defects, and it is up to the user to decide what level of defect is unacceptable. In some mass production industries very high defect levels are acceptable (21).

Visual inspection, and visual inspection with aids such as dye-penetrant or magnetic particle techniques, are important, particularly during laboratory trials. They are cheap to use and can be applied anywhere; however, they can detect flaws at the surface only.

Radiography is a very useful technique for detecting all kinds of defects within a weld and has been very widely used on laser welds, particularly in laboratory trials. However, it requires strict safety precautions and some time to process and interpret the results, so it is not ideally suited to use on the production line.

Ultrasonic detection is the most sensitive technique and can detect defects too small to be found by other methods. However, it requires a high level of skill in interpretation and has not been used much with laser welding. Ultrasonic detection provides data very quickly, which makes it potentially very attractive for incorporation into an automatic production line, but the many technical difficulties associated with this have still to be overcome.

Some applications lend themselves to novel NDT techniques, or novel adaptations of existing techniques. For example, Salfi and Vahaviolos (22) have proposed the use of stress wave emission to evaluate in real time the pulsed laser welding of wires to terminal posts. Such novel techniques are not in widespread use, however, and their usefulness remains to be proven.

Destructive Testing

One of the most widely used destructive tests is the metallographic section. A section is cut from the weld, mounted, polished and etched for examination under an optical microscope. A typical procedure for the preparation of a section of a laser weld has been described in Reference 23. Sections of welds can be used as a simple check to determine the profile of the fusion zone or can provide detailed metallurgical data to explain the existence of defects or poor mechanical properties, see Figure 10.

Many detailed laboratory studies of laser welds have been reported in the literature for a great number of materials, such as aluminium alloys (24), structural steels (25,26), stainless steels (26,27), and dissimilar metals (28). These report data derived from sections and simple mechanical tests such as bend, tensile and impact tests. Bend and tensile tests are relatively cheap and easy to use, so can form part of laboratory trials, or can be used in batch testing of production welds to ensure adequate strength, ductility and freedom from defects. When carbon steels are welded the laser weld can be much harder and stronger than the parent material because of the rapid cooling rate experienced by the weld. This can make interpretation of these tests in laser welded carbon steel more difficult than normal, as deformation and failure almost always occur in the parent material, not the weld, but even so a useful check of weld quality is obtained.

As section thickness increases brittle fracture is a greater risk and so impact properties become important. The results of Charpy and dynamic tear tests in laboratory trials are reported in the literature, e.g. References 26 and 29, but such tests are not widely used in production. Good properties in Charpy and dynamic tear testing are reported but, as has been pointed out by Goldak and Nguyen

(30), these tests may be misleading in cases where the crack path does not follow the weld but is diverted into the parent metal. As the weld is much stronger than the parent material this can happen, even though the weld is more brittle than the parent material.

The properties of welds under cyclic loading, i.e. in fatigue, are particularly important in some applications but very little work appears to have been done on the fatigue properties of laser welds, although some work has been reported on low alloy steels and titanium alloys (31).

REFERENCES

1. Courtney, C. and Steen, W.M. 'Measurement of the diameter of a laser beam', Applied Physics (Germany), 17 (3), November 1978, pp. 303-307.
2. Meyerhofer, D. 'Measurement of the beam profile of a CO_2 laser', IEEE J. Quantum Electronics (USA), QE-4 (11), November 1968, pp. 969-970.
3. Foulk, L.R. 'Laser beam profile and detector surface scanner', Report DBX-613-1841 (Rev), Publ. Kansas City, Mo, Bendix Corp., November 1978, p.32.
4. Grosjean, D.F., Olson, R.A. and Sarka, Jr, B. 'High power CO_2 laser beam monitor', Rev. Sci., Instrum (USA), 49 (6), June 1978, pp. 778-781.
5. Sepold, G., Juptner, W. and Rothe, R. 'Remarks on deep penetration cutting with CO_2 lasers', in Welding Research in the 1980's, Proc. Int. Conf., Osaka, 27-29 October 1980, Publ. Osaka University Welding Research Institute. Addition Paper A-29, p.3.
6. Crafer, R.C. and Oakley, P.J. 'Review of continuous wave CO_2 laser beam measurement techniques, and the development of a high intensity beam scanner', Welding Institute Research Report 165/1981, October 1981.
7. Lim, G.C. and Steen, W.M. 'Measurement of the temporal and spatical power distribution of a high power CO_2 laser beam', Opt. and Laser Tech. 14 (3), June 1982, pp. 149-153.
8. Sanderson, A. 'Electron beam monitoring technique and probe trace analysis', Welding Institute Research Report Misc. 34/10/75.
9. Dowden, J., Davis, M. and Kapadia, P. 'The flow of heat and the motion of the weld pool in penetration welding with a laser', J. Appl. Physics, Vol. 57, pp. 4474-4479, 1985.
10. Dowden, J., Davis, M. and Kapadia, P. 'The molten region temperature distribution in laser welding', J. Phys. D. (Applied Physics), Vol. 18, pp. 1987-1994, 1985.
11. Dawes, C.J. 'An introduction to CO_2 laser welding low carbon steel up to 4mm thick', Paper 43, Proc. Conf., Developments and Innovations for Improved Welding Production, Birmingham, 13-15 September 1983.
12. Jimbou, R. et al. 'Fusion characteristics in CO_2 laser welding', Proc. Int. Conf. Welding Research in the Eighties, JWRI, Osaka, 1980.
13. Shinada, K. et al. 'Basic study on laser welding', Paper 3 in Proc. Int. Conf. Laser Processing, Anaheim, CA, November 1981.
14. Pauley, J.T. and Russell, J.D. US Patent No. 4127761, November 1978.

15. Banas, C.M. US Patent No. 4152575, May 1979.

16. Doherty, J. and Holder, S.J. 'Adaptive control or arc welding', Paper 45, Proc. Conf. Developments and Innovations for Improved Welding Production, Birmingham, 13-15 September 1983.

17. Gnanamuthu, D.S. 'Laser surface treatment', Applications of lasers in materials processing, Proc. Conf., Washington, DC, 18-20 April 1979, published Metals Park, OH44073, American Society for Metals, 1979, pp. 177-211.

18. Hella, R.A. 'Materials processing with high power laser', Optical Engineering, 17 (3), 1978, pp. 198-201.

19. Ayers, J.D., Tucker, T.R. and Schaefer, R.J. 'Wear resisting surfaces by carbide particle injection', Source book on Applications of the Laser in Metalworking, published: Metals Park OH44073, USA, American Society for Metals, 1981, Ed: E.A. Metzbower, pp. 301-309.

20. Breinan, E.M., Thompson, E.R., Banas, C.M. and Kerr, B.H. 'Assessment of advanced laser materials processing technology', United Technology Corporation Research Centre Report R77-9122887-3, 1977, 99 pages.

21. Eckersley, J.S. 'CO2 laser welding of aluminium air spacers for insulated windows', Proc. Conf. ICALEO 1982, Boston, September 1982.

22. Saifi, M.A. and Vahavlolos, S.J. 'Laser spot welding and real time evaluation', IEEE J. of Quantum Electronics, QE-12 (2), February 1976, pp. 129-136.

23. Anon. 'Checking the quality of laser weld', Weld. J., 58 (7), July 1979, pp. 53-54.

24. Moon, D.W. and Metzbower, E.A. 'Laser beam welding of aluminium alloy 5456', Weld. J., 62 (2), February 1983, pp. 53s-58s.

25. Breinan, E.M. and Banas, C.M. 'Preliminary evaluation of laser welding of X-80 artic pipeline steel', WRC Bulletin, 201, December 1974, pp. 47-57.

26. Willgoss, R.A. et al. 'Assessing the laser for power plant welding', Weld Metal Fab., 47 (2), March 1979, pp. 117-127.

27. Estes, C.L. and Turner, P.W. 'Laser welding of a simulated nuclear reactor fuel assembly', Weld. J., 53 (2), February 1974, pp. 66s-73s.

28. Seretsky, J. and Ryba, E.R. 'Laser welding of dissimilar metals: Titanium to nickel', Weld. J., 55 (7), July 1976, pp. 208s-211s.

29. Stoop, J. and Metzbower, E.A. 'A metallurgical characterisation of HY-130 steel welds', Weld. J., 57 (11), November 1978, pp. 345s-353s.

30. Goldak, J.A. and Nguyen, D.A. 'A fundamental difficulty in Charpy V-notch testing narrow zones in welds', Weld. J., 56 (4), April 1977, 119s-125s.

31. Fraser, F.W. and Metzbower, E.A. 'Solidification structure and fatigue crack propagation in EB welds', Proc. Conf. Applications of Lasers in Material Processing II, January 1983, Los Angeles.

(a)

(b)

Figure 1. Carbon dioxide laser equipment
    a) 1200 watts, Courtesy Electrox Ltd.
    b) 2000 watts, Courtesy Control Laser Ltd.

74

Figure 1. continued.
　　　　　c) 10000 watts, Courtesy Ferranti Industrial
　　　Electronics Limited.

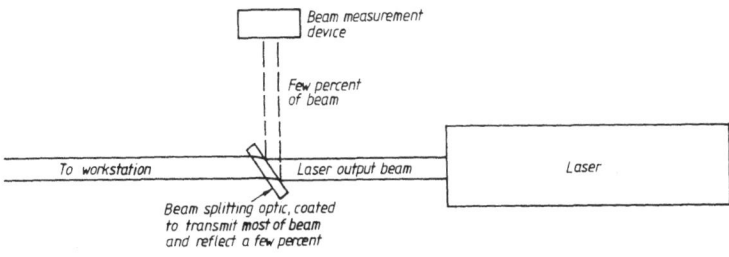

(a)

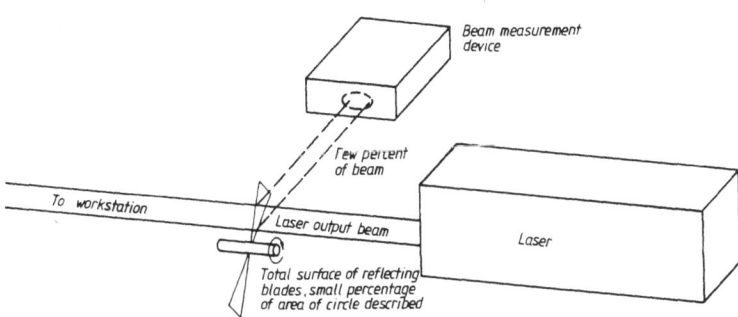

(b)

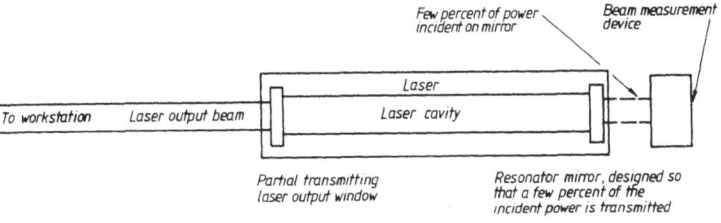

(c)

Figure 2. Methods of sampling laser beam.
     a) Static beam splitter
     b) Rotating chopper mirror
     c) Via laser resonator mirror

Figure 3. Water-cooled calorimeter.

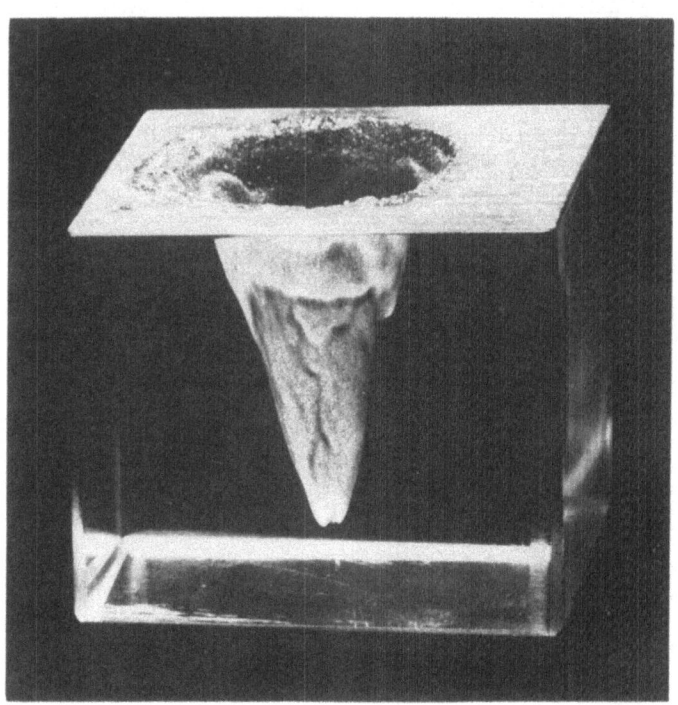

Figure 4. Deep burn print in acrylic material.

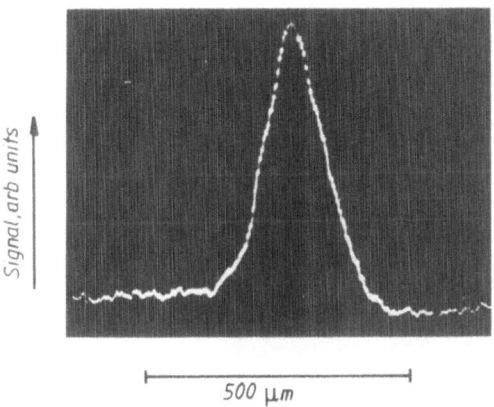

Figure 5. Typical oscillogram from high intensity beam scanner.

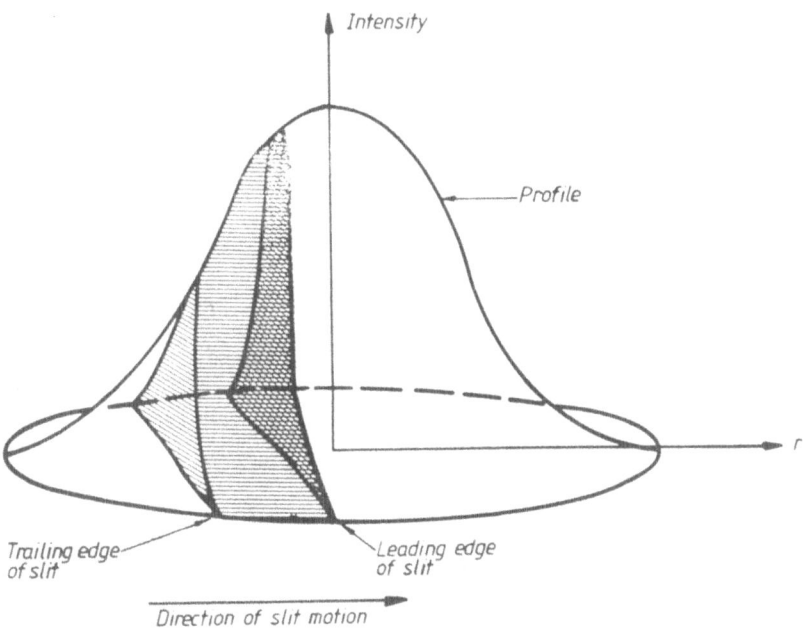

Figure 6. Geometry of beam scanner interaction. Shaded volume is proportional to detector signal. The three dimensional shape is the intensity distribution. The two dimensional curve through the axis is the intensity profile.

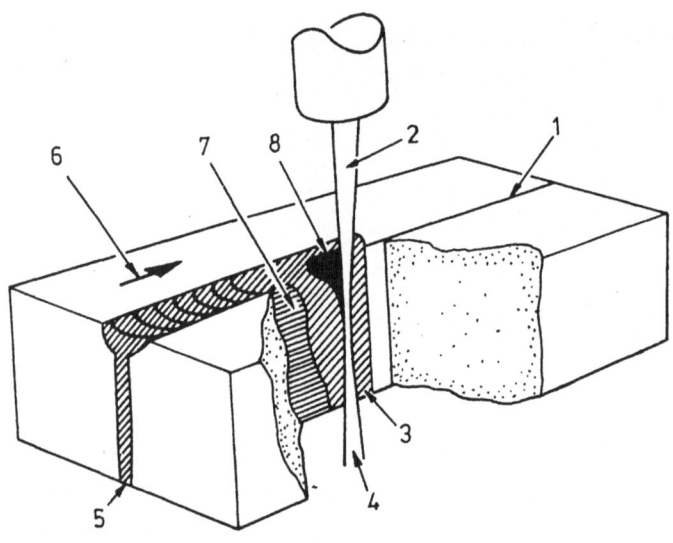

Figure 7. Deep penetration keyhole. 1) Close butt joint. 2) Laser beam. 3) molten metal. 4) Proportion of power passes through keyhole. 5) Full penetration weld. 6) Welding direction. 7) Solid weld bead. 8) Keyhole.

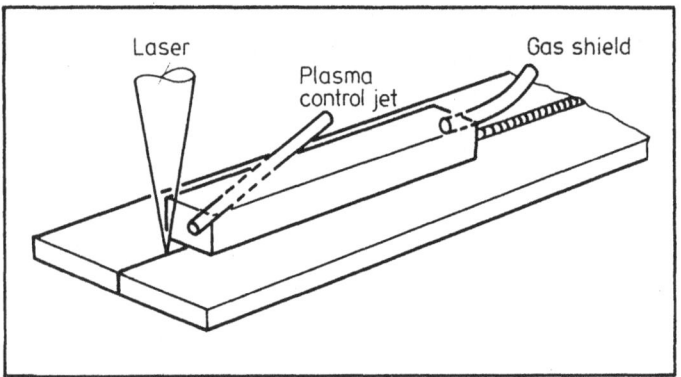

Figure 8. Schematic of plasma control device.

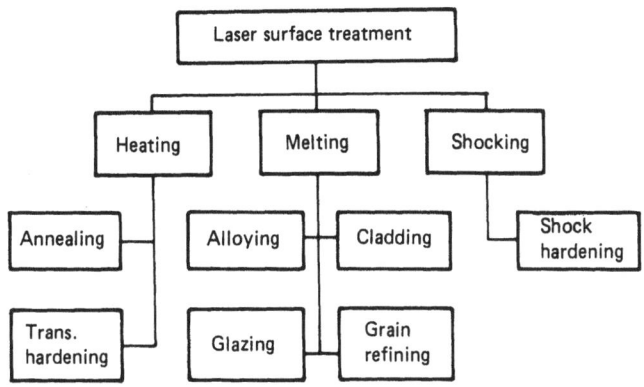

Figure 9. Methods of laser surface treatment of materials.
(from reference 17)

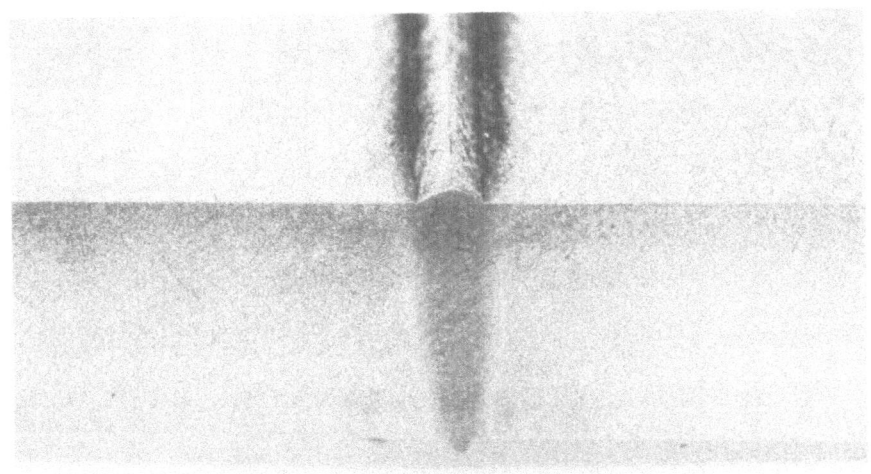

Figure 10. Metallographic section of laser weld.

NON-METALLIC MATERIALS PROCESSING: AN INTRODUCTION

I.J. SPALDING

UKAEA Culham Laboratory, Abingdon Oxfordshire OX14 3DB, UK

1. SOME OPTICAL BEAM CONSIDERATIONS

As discussed in standard texts (1-3), the output from a lowest-order mode laser operating with a stable optical resonator has the Gaussian intensity distribution.

$$I(r) = (2P/\pi a^2) \exp (-2r^2/a^2) \qquad (1)$$

where r is the radial distance from the beam centre, P is the laser power, and a is the $1/e^2$ intensity point. This "TEM_{00}" beam maintains its shape (but changes its scale) as it propagates along the Z axis in the following way:

$$a^2(z) = a_0^2 \left[1 + \left[\frac{\lambda z}{\pi a_0^2}\right]^2\right] \qquad (2)$$

where a_0, the 'beam waist', is a function of the mirror curvatures and separation. For trouble-free machining a_0 should be constant and independent, for example, of laser power. It follows from this 'ideal-laser' resonator theory that if the collimated beam is focused by aberration-free optics, the focused spot diameter

$$D \simeq 1.27 \ F\lambda = 2\lambda f/\pi a \qquad (3)$$

where F(>1) is the F number, defined as the focal length (f)/incident beam diameter (2a), and the physical size of the focusing aperture significantly exceeds 2a (to avoid truncation of the Gaussian beam).

The depth of focus (to half peak intensity) $Z_{0.5} = \pm \dfrac{4\lambda F^2}{\pi}$ (4)

The practically more useful 90% intensity point $(Z_{0.9})$ can be calculated from the relation $3Z_{0.9} = Z_{0.5}$. In a 'perfect' laser the depth of focus, and the spot diameter will thus both be larger at longer wavelengths. (Quantitatively, $Z=13.5 \ F^2$ μm and $D=13.5F$ μm for $\lambda=10.6$μm radiation.) In general a large depth of focus but a small focal spot size are advantageous for typical non-metal applications such as cutting, so that some design compromise in the F-number of the focusing system is required. It is usually easier to extract more power from a laser when operating with the higher order transverse resonator-modes (1-3), but this is at the price of a worsened spot-size and depth of focus. For welding and heat-treating, the materials-science of the workpiece and the interaction of the laser-beam with emergent vapours or plasmas, especially for metals, may favour these alternative stable-resonator intensity distributions. The same general design compromises apply to the more complex amplitude and phase

TABLE 1. Energies needed to melt* or vaporise some common materials. (Source for non-metals: Ref.6).

MATERIAL	Q(GJ/m)
Aluminium	~ 2.5*
Plywood	7.9
Plexiglass	7.9
Tungsten	~12.5*
Fibre glass epoxy composite	36
Concrete	42
Vaporisation of most metals	30-80
Boron epoxy composite	69
Glass	78

distributions exhibited in the outputs from unstable resonator lasers, which can also generate highly directional - and so readily focused - beams (3). (Industrial lasers employing unstable resonators are most commonly short and 'fat'. That is, they are low Fresnel-number systems: such as transverse-flow CO_2 lasers which generate average powers in a range from about 1 to 20kW - powers more generally identified with processing metals rather than non-metals).

2. THERMO-MECHANICAL FACTORS

The energy per unit volume (Q) needed to melt or vaporise a few common materials is summarised in Table 1. The incident laser energy will - of course - need to be higher, in order to compensate for losses due to transmission, reflection, conduction, and occasionally plasma formation at the workpiece. For non-metals transmission losses can sometimes be significant (most obviously, through glasses at visible wavelengths and through thin plastic sheeting in the infra-red). Reflection losses are not usually very important for non-metals, although they will often dominate the initial (room-temperature) coupling to clean, unoxidised, metallic targets (cf Table 2). A convenient estimate of the importance of conduction losses

TABLE 2. Common industrial lasers, operating wavelength (λ), and reflectance.

LASER	TYP. POWER kW (MEAN)	λ (μm)	REFLECTANCE[1]				
			Au	Ag	Cr	Ni	SiO_2
CO_2 gas	0.01-20	10.6	0.98	0.99	0.98	0.98	0.1-0.3[2]
Nd-YAG	<0.1	1.06	0.98	0.96	0.90	0.90	
Argon-ion	0.02	0.488	0.42	0.95	0.44	0.44	
KrF(pulsed)[3]	<0.1	~ 0.249					

NOTES
1. Room temperature, normal incidence (i.e. polarisation-insensitive angle) unless specified.
2. 0.1 at 15° angle of incidence, 0.3 at 70° angle of incidence - this variation can be important for hole-drilling etc.
3. Typical application photo-lithography.

can be made by using the one-dimensional time-dependent solution of the heat-diffusion equation (4-5). For a uniform-intensity laser beam absorbed for a time τ on a thin layer (very typically $<\lambda$ in depth) on the flat surface of a semi-infinite medium, the depth at which the temperature rise is half that at the surface is given by

$$X = 0.68 \ (K\tau)^{\frac{1}{2}} \tag{5}$$

where K is the thermal diffusivity.

Table 3 provides a few illustrative numbers: it will be seen that on the time-scales of practical interest thermal conduction is important (amongst other heat-transport mechanisms) for the CW metal-working applications to be discussed in later lectures, but is normally unimportant for non-metal working.

The machining speed (V) for vaporisation of a cut, or "kerf", width W ($\sim$ D) in a material of thickness (t) is consequently often given by the simple relation

$$V = P/QWt \tag{6}$$

Some purely illustrative speeds are given in Table 4. (It should perhaps be noted that the absorption of light is polarisation-dependent on some, and more especially metallic, surfaces - so that both the cutting speed and shape of cut can then be sensitive to the relative orientation of the electric vector and the cutting direction, cf. Fig 1. Where this might prove a problem, it is usual to 'average-out' the effect, e.g. by using a circularly polarized laser beam or by mechanically rotating the electric vector of a linearly-polarized beam using appropriate mirrors). More detailed numerical modelling is rarely undertaken for this type of work (6-8) - possibly because the relevant data base is less extensive than for metals and the material forming the workpiece can also be less homogeneous and reproducible. (It is therefore particularly important to establish

TABLE 3. Heat conduction - orders of magnitude (i.e. $X^2 \sim K\tau$).

MATERIAL	(1D) THERMAL PENETRATION TIME (T/T_0=0.5 ISOTHERM)		(1D) SURFACE MELTING TIME (POWER DENSITY ABSORBED=F_0)	
	X=0.1cm	X=1.0cm	$F_0=10^4$W/cm^2	$F_0=10^6$W/cm^2
Al	0.012(sec)	1.2(sec)	0.04(sec)	4 x10^{-6}(sec)
Cr	0.056	5.6	0.14	1.4x10^{-5}
Ti	0.14	14	0.014	1.4x10^{-6}
soil, concrete 4)	2.4	237	-	-

4. This number is also representative of materials such as fused silica etc. Plastics such as polyethylene can have a thermal diffusivity an order of magnitude smaller (and have a penetration time up to 10 times larger). Table 2.3 of Ref.6 is in error, with conductivities and diffusivities transposed.

TABLE 4. Illustrative cutting speeds, using CW CO_2 lasers 5)

MATERIAL	THICKNESS (mm)	SPEED (m/s)	KERF (mm)	POWER (kw)
Paper	Newsprint	>10	0.13	0.4
Bond paper	0.05	5	-	0.25
Mylar	0.025	> 5	0.15	0.3
Textile	(0.45gm/m²)	0.83	0.25	0.4
Leather	3.0	0.05	-	0.22
"	1.6	0.3	-	1.0
Plywood	18	0.02	-	1.0
"	20	0.005	-	0.225
Glass fibre	1.5	0.05	0.25	0.4
(Resin bonded)	1.6	0.25	-	1.0
Silica	2.0	0.017	0.25	0.4
"	3.1	0.012	-	0.5
ABS plastic	0.25	10	-	0.375
"	6.3	0.10	-	1.0
Acrylic plastic	6.3	0.083	-	1.0

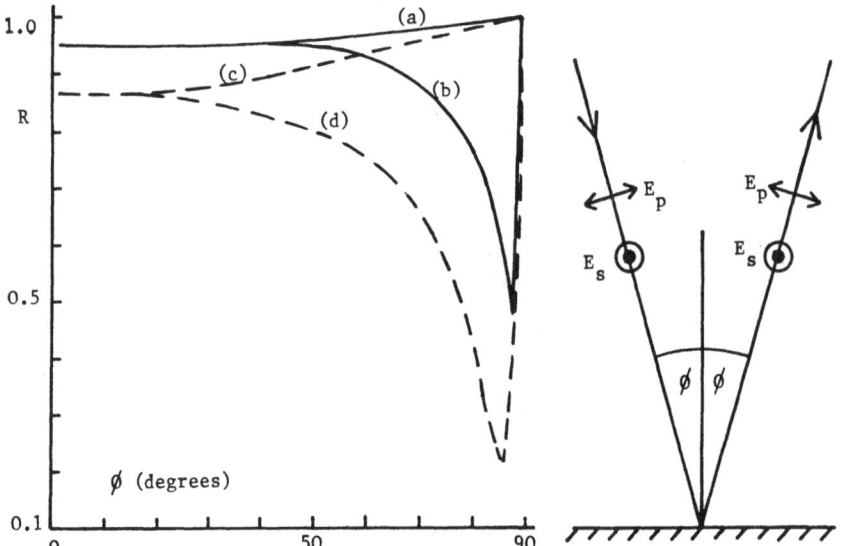

FIGURE 1. Variation of reflectivity (R) with angle of incidence ϕ for copper at room temperature (a), (b) and 1000°C (c), (d). Curves (a) and (c) are drawn for 10.6μm radiation linearly polarized with the electric-vector E_s parallel to the surface of the workpiece; and (b), (d) with the orthogonal polarization (E_p).

5. Where kerf width is stated, source is Ref.5 and Culham Lecture Note CLM-L21 (Lectures by Spalding at International School of Quantum Electronics, Erice, Sicily - June 1973). Other data is abstracted from Ref.6 and recent trade catalogues.

experimentally the process tolerances of interest before major investment decisions are made). However, these very basic ideas and formulae can prove - with some important qualifications indicated above - to be extremely useful in practise. In particular a sound understanding of laser optics is essential for any systems-designer or maintenance-engineer wishing to efficiently and reliably pipe a laser-beam through a beam-delivery system onto a workpiece, even if he is prepared to treat most of the other aspects of the laser he has bought as a 'black box' - merely providing him with a highly controllable and localized heat (or light) source.

3. WHY USE A LASER?

The capital cost of a laser is not insignificant (although it is usually smaller than the cost of mechanical and NC (numerically controlled) equipment installed with it on a typical production line) so it is instructive to tabulate some of its potential advantages:

(i)   The cut, or hole, can be very narrow. (Material savings, less dust?)
(ii)   The cut, or weld, can have a narrow heat-affected zone, HAZ (giving low distortion).
(iii) There is no 'tool drag' - only a need to locate the workpiece.
(iv)   Therefore, 'tool wear' is minimized.
(v)   Most thin materials can be cut, (whether hard, soft, brittle, friable etc.)
(vi)   Selective cutting is possible (e.g. stripping insulation from wires; semi-conductors etc.)
(vii) Light is inertialess: starting, stopping, turning corners etc.can be accomplished relatively easily at high speed. (α-numeric displays, bar codes, plywood die-board manufacture etc.)
(viii) Multistation working over long distances is possible. (Culham has already tackled 70m).
(ix)   Laser machining is NC compatible, and 'tooling changes' are relatively easy. (Gas shrouds, for welding, possibly pose the greatest problem).
(x)   It is often possible to weld without special edge preparation - e.g. immediately after a laser-cut. (Simultaneous cutting and welding of twin-ply plastics have been achieved).
(xi)   There can be significant environmental advantages (e.g. less noise, dust or swarf).
(xii) The nature of the cut edge may sometimes give it an advantage over an alternative technique: e.g. it may be squarer or smoother than a bandsaw's; show less mechanical stress; be sterile (for medical applications); or the workpiece may show less relaxation and change of shape after cutting (e.g. rubber).

This check-list is meant to be illustrative rather than exhaustive. Table 5 details some specific processes - many of which are already finding practical, i.e. economic, application in industry. (The processing of a very wide range of non-metals was illustrated using colour slides/videos for this introductory lecture but only a few are reproduced here, in order to avoid repetition with the subsequent, specialist, articles in this volume).

Several very detailed technical and economic case-studies of the potential for NC laser cutting have been published. In one particular study, for example, seven alternative methods for profile-cutting sheet metals, acrylic plastics and plywood were compared. For the particular profile of mixed component sizes and compositions appropriate to the

TABLE 5. Practical laser applications (*illustrated in video/slide)

Process	Typical Application	Laser
Laser Cutting		
i. Controlled, thermal fracture	Test tubes* 0.7mm Al_2O_3 - 1m/min Single crystal quartz Ferrite, sapphire etc.	CO_2 (5-25W)
ii. Melt & blow	Glass*	CO_2
iii. Reactive gases	O_2 on metals etc.	CO_2
iv. Vaporization (~$10^{12}W/m^2$) (CW on non-melting materials or pulsed lasers)	Graphite Si or B nitride Silica* printing*, leather*, plywood*	Nd-YAG Nd-YAG CO_2 CO_2
v. Scribing (blind holes; stress raisers etc.)	Si	Nd-YAG (kHz)
vi. Drilling (successive pulses on same hole; blasting - $H2_0$ on Cu)	Wire dies, video discs, ceramics 100:1 aspect ratio)	Pulsed CO_2, YAG Excimer
Laser Welding (NB - metal welding of electronic components through glass, using visible lasers)	slit/seal plastics* optical fibre splicing optical flats (sealing)	500W CO_2 2W CO_2 <100W CO_2
Surface Treatments		
i. Cleaning	Atomically clean Si	$2J/cm^2$ ruby
ii. Crystal growing (floating zone)	Y_2O_3, Al_2O_3, pyrolytic C	CO_2
iii. Deposition	ZnSe (1nm s^{-1}) Cr (10μm s^{-1})	CW CO_2 YAG
iv. Chemical vapour deposition	Si from SiH_4 (100nms^{-1})	50W CO_2
v. Photo etch/deposition (Tune λ?)	CH_3Br/GaAs, InP Cd on SiO_2	Ar^+ Ar^+ x2
vi. Marking (ablation)	Bar codes on coke tins	TEA CO_2

company (9), a laser-cutting system was preferred to a mechanical 'nibbler' system: a 2mx3m gantry system was used to move the (relatively light) focusing optics at accelerations of up to 1g and speeds of up to 40mmin^{-1} and a typical cutting speed of 4.5mmin^{-1} (determined by the available laser power) was achieved on 20SWG (0.914mm thick) sheet.

Laser marking techniques are now also very specialized (10), e.g.:

(i) CW CO_2 lasers are often scanned over a reflective mask (in close proximity to the workpiece) to provide very high detail reliefs on wood, stone, rubbers etc. (This rubber may then be wrapped around a roller, for the flexographic printing of wall papers etc.)

(ii) Pulsed CO_2 lasers are often used to illuminate a complete mask (in a pulse lasting only a few μs) to provide single-shot 'best-before' or bar-code marking of packaging; this is most commonly applied to non-metals, and with suitable relay-optics can be used on moving packages.

(iii) Pulsed Nd-YAG lasers are used in a dot-matrix mode to produce deeper marks, e.g. to register part-numbers on (metallic) aeroengine components. For example, an acousto-optically switched Nd:YAG laser operating at 15kHz and generating a 7x5 spot matrix can mark moving cables, tubing etc. with between 306 and 750 alpha-numeric characters per second (11).

(iv) CW Nd-YAG lasers can 'engrave' high quality marks (such as trade marks), using galvanometer mirrors to deflect the beam onto the product. This technique proves well suited to 'difficult' marks ideally requiring continuous script, such as Arabic or Chinese characters (10).

4. SYSTEM REQUIREMENTS, AND TRENDS
4.1. Lasers

Market surveys (12,13) show that most conventional machining and marking operations are currently undertaken using pulsed or CW Nd:YAG(λ=1.06μm) or CO_2(λ=10.6μm) lasers, within the mean power range of a few watts to a few hundred watts (Nd-YAG) or 1~2kW (CO_2). These lasers have electrical efficiencies typically in the range $<$2% (Nd-YAG) and 7-15% (CO_2). Such systems have a long pedigree: they have proved relatively compact, flexible, and reliable, and the industrial versions are becoming increasingly refined.

In particular, a wide range of axial flow CO_2 lasers, producing Gaussian or quasi-gaussian mode outputs are now available at powers of up to ~1.5kW ('slow flow) and ~5kW ("fast", i.e. near sonic, flow). At higher powers it is often convenient to cool the laser medium perpendicular to both the discharge and optical axes, and transverse-flow CO_2 lasers of this generic type are now commercially available in the power range 1~20kW. Similarly, solid state Nd-YAG (or ruby) systems often provide sophisticated monitoring and control of flash lamp performance (over lives of typically many millions of shots), and may offer shot-to-shot pulse-shape control (via the charging cycle).

For these infra-red lasers any decomposition of the workpiece is normally thermal. Shorter wavelength visible (argon ion, copper vapour etc.) and UV (ArF, λ=193nm to XeF, λ=351nm) lasers are being used increasingly for lower mean-power applications where a smaller spot-size, photo-chemical, bond-breaking or other specialised absorption effects are required. (An extreme example is provided by the very wide range of medical applications, which will not be addressed here).

4.2. Beam Delivery Systems

Visible and 1μm wavelengths are now often handled using fibre-optic (flexible optical waveguide) techniques. (For example, JK Lasers recently exhibited multi-station working via multiplexed CNC controlled fibre-optic delivery systems at the Laser '85 exhibition in Munich). It is worth noting that the final accuracy of the machining operation is usually determined by the mechanical accuracy of work and beam handling components. For example, printed circuit boards are routinely ruby laser-drilled to tolerances of ±2.5μm by Advanced Laser Systems, of Massachusetts (14). Multikilowatt 10μm (CO_2) wavelengths are more conveniently handled using lenses or mirrors, and with care a beam which is almost diffraction-limited can then

be delivered to the workpiece. The repetition accuracy of any mechanical
device will in general depend on the weight,mechanical dimensions and
speed/accelerations required of the moving components, and this topic will
be discussed in greater detail in subsequent lectures. However, as a
general guideline, for systems in which a relatively light multi-axis laser
focusing head is moved over the workpiece of several m^2 dimensions (rather
than moving the laser itself, or the work), the following accuracies might
typically be expected:

(i)	Gantry	$\sim\pm$ 0.05mm
(ii)	NC (numerically controlled) XY table	$\pm$ 0.1 mm
(iii)	Industrial robot (5 axis)	$\pm$ 0.2 mm
(iv)	Optical following system	$\pm$ 0.3 mm

Fig. 2 illustrates one of the very first (5 axis) industrial laser-robots.

4.3. Auxiliary Equipment

It should always be remembered that the focusing head often has to carry
ancillary equipment, such as gas shrouds or cutting jets, position sensors
etc. Each of these components must be fully compatible with the other
systems-requirements of paras. 4.1 and 4.2. In addition, they may have an
important and independent influence in the general reproducibility of the
process. Culham has, for example, investigated the many factors influencing
the reproducibility of the pressure exerted on the workpiece as the cutting
nozzle design is varied, or the gas supply is adjusted under conditions
appropriate to supersonic flow, cf. Figure 3. For practically-convenient
stand-off distances (i.e. at least a few mm between the nozzle and the
workpiece) it is important to choose design parameters which permit a

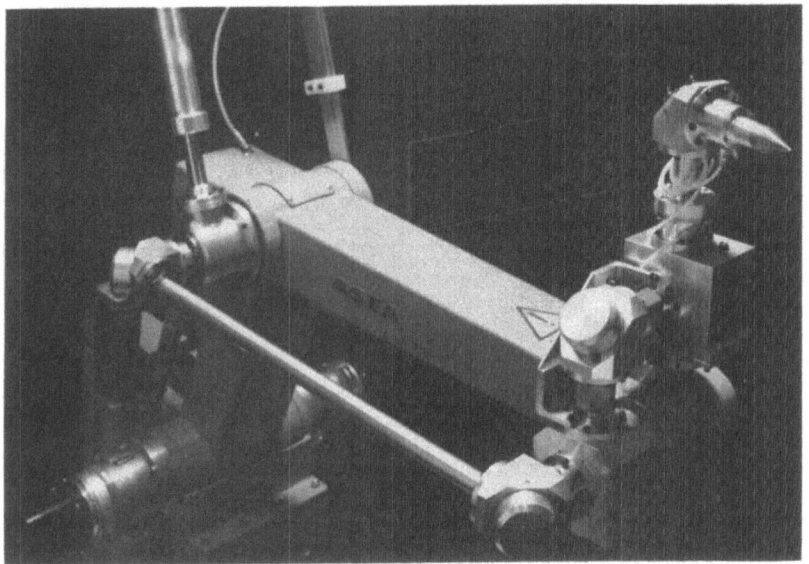

FIGURE 2. COBRA, a flexible $\frac{1}{2}$-1kW laser beam guide coupled to an ASEA 1RB6
industrial robot, developed at Culham for FLS Ltd, now a division of
Ferranti plc. (Systems such as this will cut a wide range of plastics and
metals with a repetition accuracy better than ±0.1mm).

reasonable process tolerance. Finally, the ability to monitor the focal
intensity distribution (Fig. 4), and by inference the process heat
distribution, on-line (16) is potentially a very valuable feature of
laser-processing since this offers the prospect of closed-loop control.
(Such control could be applied, for example, to the laser resonator - to
stabilize incipient variation of spatial mode, or directly to the process
itself, via a suitable control algorithm).

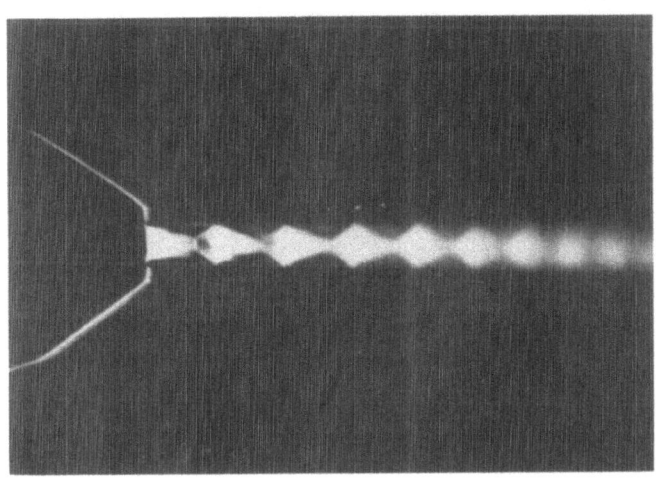

FIGURE 3. Schlieren photograph of the stationary pressure wave from a
supersonic cutting jet (17).

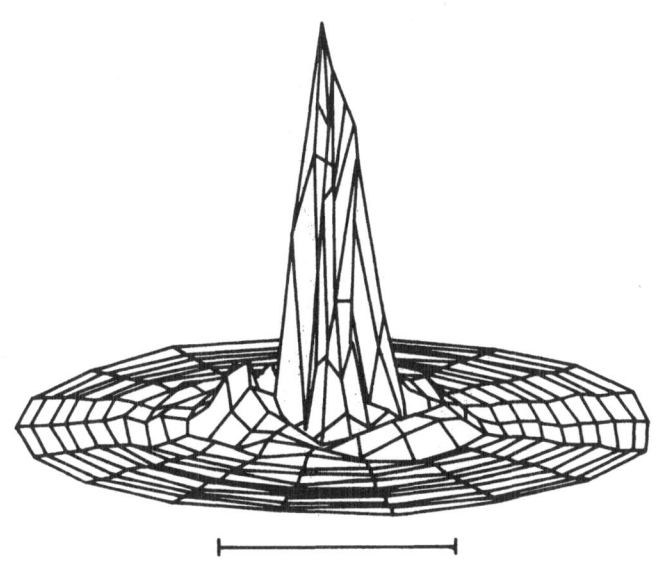

FIGURE 4. Fully two-dimensional ON-LINE measurements of the focal intensity
distribution from a high-power CO_2 laser (16). The marker shows an angular
scale of 1 milliradian.

5. CONCLUSION

Technical feasibility must never to confused with economic feasibility. The example, newspoint and bond-paper can certainly be laser cut, with a dust and char-free edge, at linear speeds comfortably exceeding 600m/min and double layers of 270g m^2 polyester/wool fabrics at >60m/min (Fig. 5): but this may not always provide sufficient reason to install a system in a factory. On the other hand, the superficially more 'routine' job of die-board manufacture has been successfully exploited since about 1971 at William Thyne, as a results of pioneering work by BOC and Ferranti plc; here the laser cutting speed is relatively slow - perhaps less than 1.2m/min for 18mm plywood - but the advantages enumerated in paragraph 3 are sufficient to make it a winner. Similarly, the excellent edge quality achieved when cutting acrylic plastics (Fig. 6) has lead to a strong growth in applications for the sign-writing industry. Almost identical considerations also apply to the cutting (or sealing) of fused silica products (Fig. 7), and of thermo-plastics materials. (One of these systems has worked very successfully for 3 shifts per day, non-stop for 6 weeks at a time between routine servicing. The availability of suitable NC controls has lead to increasing commercial exploitation of the laser cutting, of carpets, leather and other materials for the automotive and similar (high-volume) industries - and even in some aspects of bank-note production! (The uniquely-contoured security thread embedded in the £50 note was produced using computer-controlled laser equipment developed and built at Culham for the Bank of England).

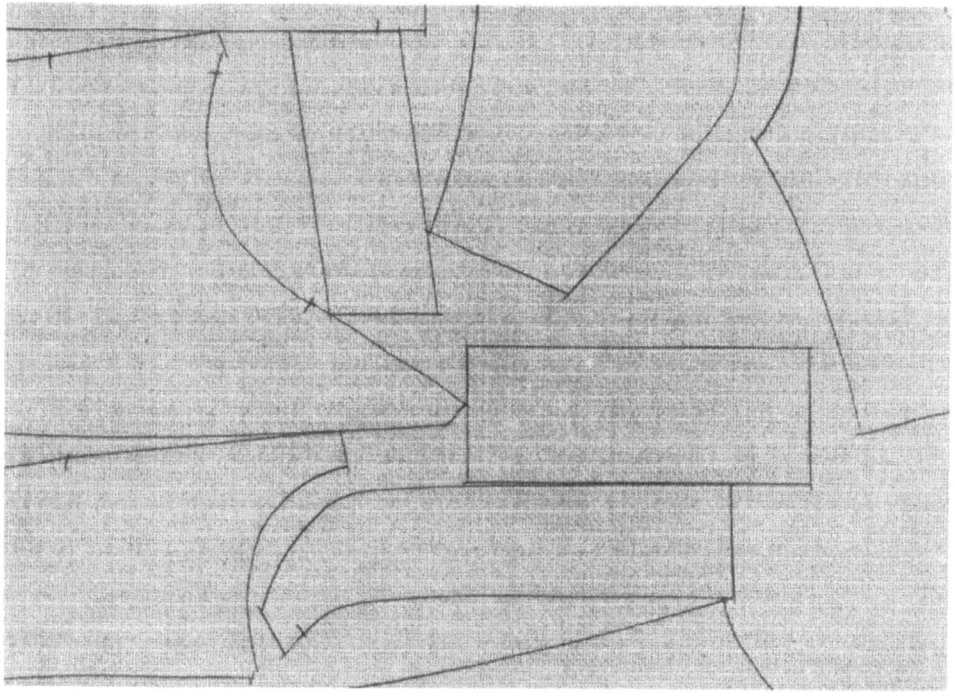

FIGURE 5. Two superimposed layers of woollen worsted fabric, laser-cut using a computer-controlled 400W CO$_2$ system. (This computer also generates the pattern, and optimizes its layout, on the patterned suiting).

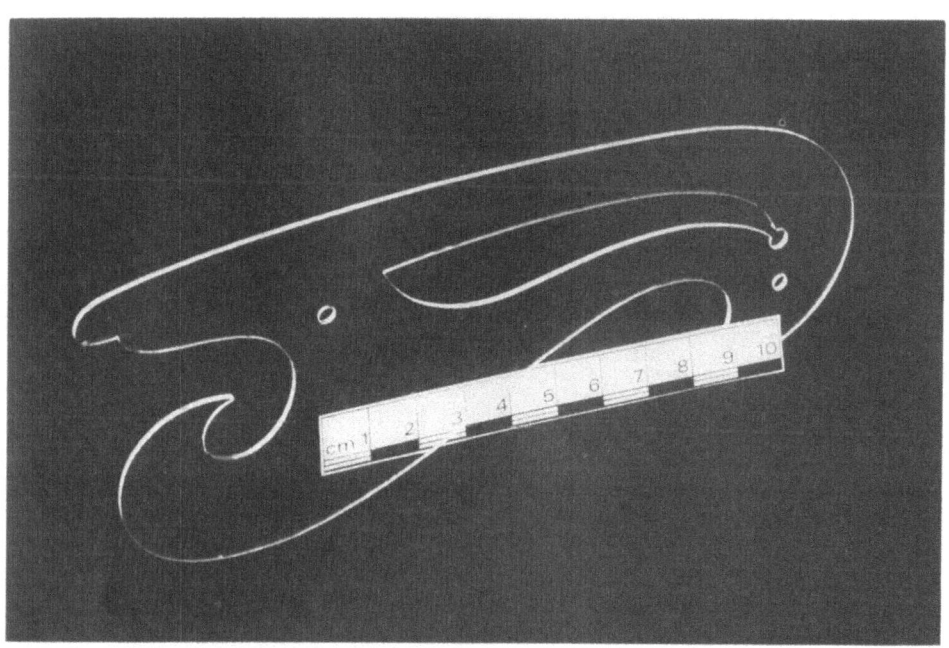

FIGURE 6. CO_2 laser-cut acrylic plastic.

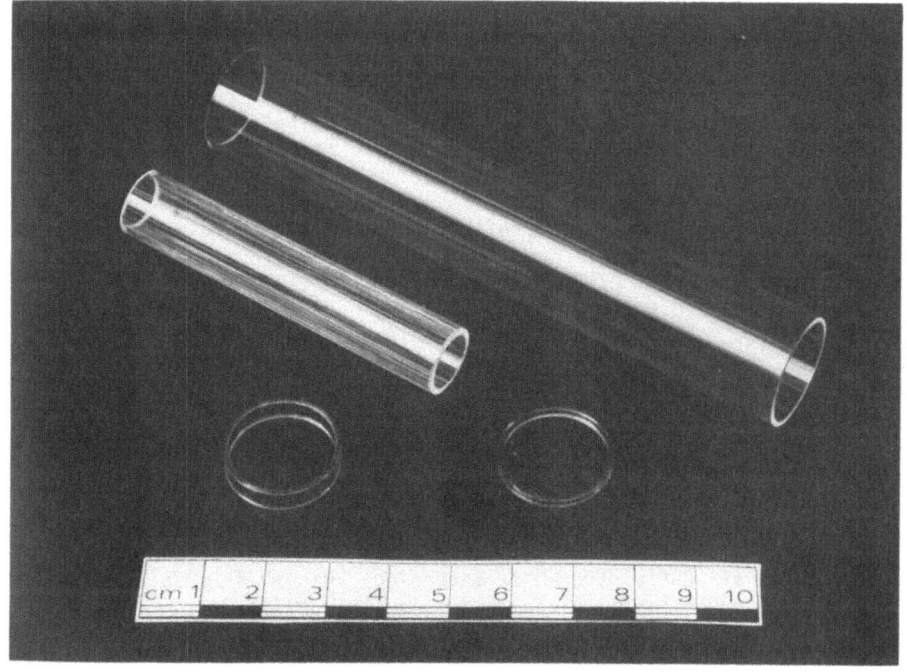

FIGURE 7. Fused-silica tubing, cut using a 400W CO_2 laser.

FIGURE 8. Etched pattern on wood, produced by rasterng a CO_2 laser over a disposable contact mask. (The potential delicacy of the effect is best illustrated with colour photography on finer-grained woods, glasses or slates etc.)

The excellent spatial resolution achievable using laser techniques has facilitated very unusual etching effects on a wide range of materials including woods (Fig. 8), ceramics, rubbers and glasses: these masking techniques are thus applied in the production of flexographic (printing) rollers. Finally, in high quality colour printing it is possible to scan the original picture and use the reflected signal to modulate a high power (engraving) laser in order to produce rotogravure cylinders by a fast, 'dry' technique: this provides us with a very practical example of "intellgent" machining. Many more applications, in both the metal and non-metal fields, are discussed in subsequent sections of this book. It should, however, already be clear why the laser is now regarded as an exciting new 'tool' for industry.

REFERENCES
1. Maitland A and Dunn MH: Laser Physics (North Holland, Amsterdam) 1969.
2. Svelto O: Principles of Lasers (Plenum, 2nd Edition, New York) 1982.
3. Spalding IJ: Physical Processes in Laser Materials Interactions, (Ed. Bertolotti M: Plenum, New York) 1983.
4. Hill JW et al: Optics and Laser Technology, Dec 1974, p276
5. Spalding IJ: Optics and Laser Technology, Dec 1974, p263
6. Duley WW: Laser Processing and Analysis of Materials (Plenum, New York) 1972.
7. Charschan SS(ed): Lasers in Industry, (Van Nostrand Reinhold, New York) 1972.

8. Bass M(ed): Laser Materials Processing (North Holland, Amsterdam) 1983.
9. Janjua MS, Rathmill K, Allen DM: Proceedings of 1st Int. Conf. on Lasers in Manufacturing, Ed. MF Kimmett, (IFS Conferences, North Holland Publishing Company) 1983, pp41.
10. Willis JB: Ref. 9, pp53.
11. Green DI, McNeish A, Harris JJ: Proceedings of 2nd Int. Conf. on Lasers in Manufacturing, Ed. MK Kimmitt (IFS Publications Ltd, Bedford, UK) 1985, pp279.
12. Laser Focus, January 1985, pp75.
13. Lasers and Applications, January 1985, pp60.
14. Rose CD: Electronics, July 8, 1985, pp49.
15. Spalding IJ et. al.: VDI Berichte 535 (VDI-Verlag, Dusseldorf) 1984, pp107. (See also Culham Laboratory Annual Report 1982, pp51).
16. Travis AJB: Gas Flow and Chemical Lasers 1984 (Institute of Physics Conference Series No. 72, Adam Hilger, Bristol 1985) pp367 and Ref 15, pp101.
17. Ward BA: to be published in Proc. of the Int. Cong. on Applications of Lasers and Electro-Optics, Boston, MA, USA, 12-15 Nov 1984. (See also Culham preprint CLM-P730).

LASER MICROMACHINING OF MATERIALS FOR ELECTRICAL AND ELECTRONIC DEVICES.

G. Rauscher
SIEMENS AG, Postfach 830953, D 8000 München 83, FRG

Introduction - The laser as a tool

Laser micromachining is only a small sector in the broad laser applications field, including

  optical communication
  information processing
  plasma generation
  spectroscopy
  holography
  isotope isolation
  metrology and testing
  medicine and
  machining of materials;

but laser micromachining for electrical and electronic devices holds a big market share in the machining-of-materials field.

The key aspect of the use of lasers as a machining tool, is the high power density of the laser light achieved by focussing with optical systems (lenses or curved mirrors):

Heating 10^4 Wcm^{-2} - 10^5 Wcm^{-2}	annealing doping hardening adjusting	Vaporization 10^7 Wcm^{-2} - 10^{12} Wcm^{-2}	trimming programming marking cutting
Melting 10^5 Wcm^{-2} - 10^7 Wcm^{-2}	recrystallization contacts for semiconductors alloy (surface) welding cutting		

Soares, O.D.D., Perez-Amor, M. (eds), Applied Laser Tooling. ISBN-13: 978-94-010-8096-5
© *1987. Martinus Nijhoff Publishers, Dordrecht.*

All of the lasers used as machining tools are of the high-power
type:

Laser	wavelength	operation	power (max.)	pulse energy (max.)	pulse frequency (max.)
CO₂	10.6 μm	c w	0 kW		
		pulse	100 W ($\bar{P}$)	2 J	2 kHz
Nd : glass	1.06 μm	pulse	50 W ($\bar{P}$)	100 J	5 Hz
Nd : YAG	1.06 μm	c w	600 W		
		pulse	600 W ($\bar{P}$)	100 J	200 Hz
		Q switch	20 W ($\bar{P}$)	50 mJ	50 kHz
Argon-ion	0.33 - 0.51 μm	c w	20 W		
Excimer	0.19 - 0.35 μm	pulse	100 W	1000 mJ	500 Hz

The highest-power CO_2 laser, which are now commercially available
for up to 20 kW output power, are too powerfull for micromachining.
In the field of laser applications for electrical and electronic
devices, only CO_2 lasers with up to 200 W output power are used.

Lasers have three different modes of operation: the typical
pulse length in "normal" pulse operation is in the "ms" range,
while the pulse length in the Q-switch mode is very much shorter;
namely in the ns range.
The modes of operation and the power levels in cw operation are
associated with typical applications:

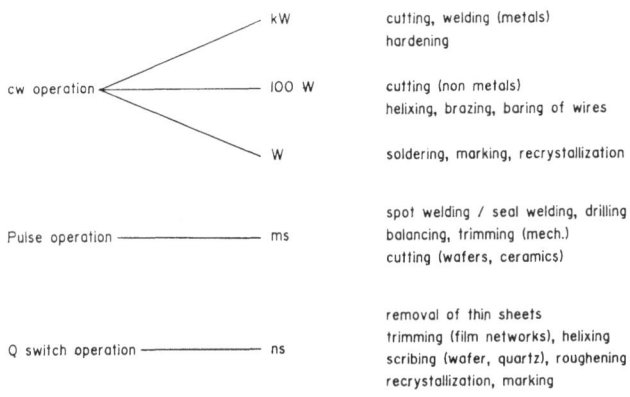

But of equal importance as the mode of operation is the wavelength
of the laser. The following table shows the correlation between
typical technical applications of lasers and their wavelengths/
modes of operation:

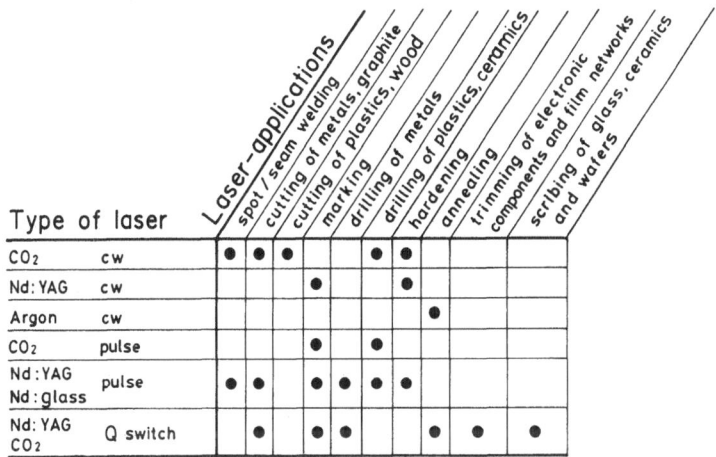

Micromachining with lasers

Lasers are widely used tools in manufacturing electrical and
electronic components. The most important areas of applications
are:

components
(electrical/electronic)
- helixing of resistors
- trimming of thin-film and
  thick-film resistors (SMD)
- cutting of wafers
- redundancy for memories
- manufacturing of PLA devices
- marking
- cleaning

electrical networks
(hybrid circuits)
- machining of ceramic substrates
  (cutting, scribing, drilling,
  marking)
- trimming (passive) and functional
  trimming of thick-film and thin-film
  networks

pc-boards
- cutting of flexible pc boards

components
(mechanical)
- relays: spot welding
         seam welding
         trimming (mechanical)
         soldering of wires

- modules for optical communication:
         fixing (adjusting) and
         spot welding
         seam welding of feed-throughs

assembly
- soldering, brazing
- welding spot welding
           seam welding
- adjustment and fixing (welding)
- trimming
- stripping

TV tubes
- spot welding
- adjustment and fixing (welding)

Lasers as high-accuracy micromachining tools have a lot of advantages, which are normally essential for laser processing in production lines for electrical and electronic components. The advantages for high-accuracy micromachining are:

- precise pulse energy

- high reproducibility

- fast control of pulse energy

- short pulse duration

- large temperature gradient

- low energy requirements

- low distortion

- no external power

- independent of atmosphere

- optical control

- combination with optical detection

- flexible tool

An additional advantage of the laser, especially for welding and drilling applications, is the high repetition rate, which in most cases is higher than the cycle time of the mechanical equipment.

For using a laser as a tool in manufacturing lines, especially
in high-volume production lines, it is necessary to choose the
optimal parameters. When using a laser in low-cost and reproducible
processes in production lines, the most important groups of
parameters for laser processing are:

MICROMACHINING with LASER

wavelength	• absorption
power	• workpiece measurements
	• station time/throughput
mode of operation	• CW
	• pulsed
	• Q-switch
quality of resonator	• distribution of power density (modes)
	in the near and far fields
	• polarization
	• stability
optical system	• power density
	• beam diameter
	• working distance

It is of great importance for every new laser application in industrial manufacturing to be scrutinized in the light of some of
the important steps of process development specified in the
following. This scrutinizing is necessary not only for completely
new applications but also for the first implementation an "old"
laser process in a given manufacturing line.
Each manufacturing line has special features; for example the kind
and design of the components produced, the production volume, the
lot sizes, the cycle time and the degree of rationalization.

The steps of process development are:

PROCESS DEVELOPMENT
general overview

● - is the problem suitable for laser processing
  - kind of laser, technical data
  - preliminary test
    for example: metallurgical problems
  - change of design, change of materials
  - choice of laser and mode of operation
    development of fundamental process
    study of process (theoretical)
  - competitive processes
    calculation of costs

● - first calculation of profitability

● - process development
        expenses
        manufacturing program
        safety
        further development of fundamental process
        reproducibility
        quality specification

● - second calculation of profitability

● - development and construction of a prototype

An alternative to the last step - development design and construction of a prototype - is the purchase of a laser system. The calculations have to be carried out by the user company, while the development and selection steps could be accomplished by a research institute, the laser system manufacture or the user company.

Laser for micromachining

Normally, production-line lasers are used in semi- or fully automatic laser machines. In most cases, a high repetition rate is
necessary due to the high investment costs.
The features of automatic laser machines are:

power network, numerically-controlled

laser parameters, numerically-controlled

machining without any contact

low costs for parts subject to wear

fast repetition rate

high investment costs

Up to now, the acceptance of the laser as a tool in manufacturing
lines has been very poor. This is due to many reasons. The expansion of laser applications could proceed faster if the following aspects were developed by research institutes, user companies
and/or laser manufacturers:

* reduction of laser manufacturing costs
    - quantity
    - modular design

* high-frequency pulsed laser
  and fast controlling systems

* controlled laser beam
  feedback control

* compact design

* flexible beam guidance
  for robots
      flexible manufacturing systems

* cooperation between
  laser manufacturers,
  the machine tool industry,
  components manufacturers
  and
  users

The main point for a highly reproducible manufacturing process
is a controlled laser beam and a fast feedback control. In
commercially available laser systems, feedback control has up to
now been confined to the positioning system. The block diagram
of an actually existing automatic laser machine looks like this:

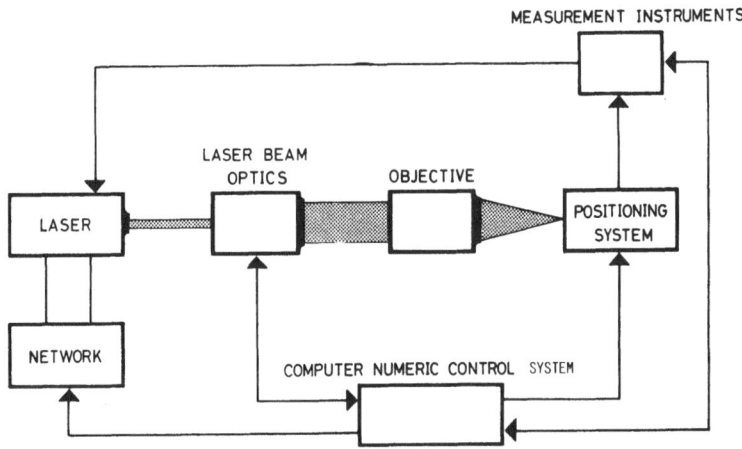

To attain more highly reproducible and more sophisticated laser
processes in manufacturing lines, especially in those for the
production of electrical and electronic devices which need very
closely toleranced laser parameters, it is necessary to develop
laser systems with feedback control capability for the interaction
between laser light and material.
The laser machines should, for example, have a diagnostic system
for the machining process and a beam monitor; the data being
processed in a computer numeric-control system modulating the laser
resonator.
A block diagram of such a sophisticated laser system would look
like this:

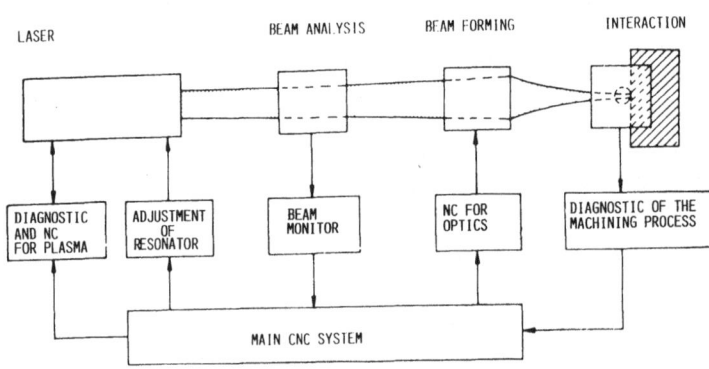

Laser applications for electrical and electronic devices

To give an overview, some special applications will be discussed
in detail:

- marking of components
- spot welding und trimming of relays
- welding of stack sheets (transformer)
- cutting of mechanical filters (material: silk)
- spot welding of fiber optic components
- cutting of silicon wafers (thyristors)
- trimming of thick-film circuits
- redundancy for VLSI memories
- laser programming of logic arrays (PLA's).

METALLIC MATERIALS PROCESSING : CUTTING AND DRILLING

by Alberto SONA

Director of the Italian CNR Program on High Power Lasers

Center for Information,Studies and Experiments,(CISE)
via Reggio Emilia N 39 - 20090 SEGRATE (MILAN) ITALY

INTRODUCTION

Cutting and drilling by lasers can be accomplished in two ways :

The first one mainly used when cutting with CW beams consists in melting the material and removing the liquid by a gas-jet. This process is often referred to as melt-cutting.

In the second case, typical of drilling and cutting by pulsed beams the material is vaporized and self-removal occurs by the expanding vapour.

In both cases the radiation - matter interaction occurs with typical features depending on the parameters controlling the process.

The most relevant are :

a - parameters depending on the laser source such as output power, wavelength, modal structure, beam divergence , beam polarization, beam diameter D.

b - parameters depending on the focusing system such as focal lenght F, f-number, aberrations.

c - parameters depending on the injected gas such as type of gas (reacting or inert), gas pressure on the target.

d - parameters depending on the target such as absorption , heat diffusivity, melting temperature, gas-reactivity.

In the following a review of the role of the various parameters will be presented.

CUTTING WITH CW OR PULSED BEAMS

The laser beam is focused by a lens in a region having a diameter d = 2.44 (λ / D) F where λ is the wavelength, F the focal lenght, D the beam diameter. The ratio f = F / D is called the f-number of the system and it has typical values between 4 and 8 . Actually, for practical reasons, the distance between the lens and the target has to be kept large enough to avoid damages to the lens. Taking into account the usual values of the beam diameters (20 - 40 mm typical) this leads to the quoted values for f. As a consequence the beam from a CO2 laser (λ =10.6 micron) is focused in a spot with a diameter of 0.1 - 0.2 mm.

For a Nd-YAG laser (λ = 1.06 micron), the size should in principle be ten times smaller. In practice,however,industrial lasers have a beam

Soares, O.D.D., Perez-Amor, M. (eds), Applied Laser Tooling. ISBN-13: 978-94-010-8096-5
© *1987. Martinus Nijhoff Publishers, Dordrecht.*

divergence ϑ which is larger then the diffraction limit; $\vartheta > 1.22 . \lambda . / D$ and the spot size is again of the order of 0.1 mm. If the beam is gaussian with a diameter D between the $1/e^2$ intensity points the spot diameter d again between the two $1/e^2$ intensity points is given by d $= (4/\pi) \lambda$ (F/D) with a depth of focus b $= (4/\pi) \lambda$ (F/D)^2 .

The spot resulting from the focusing of a gaussian beam with a multimode emission has a size increasing as the square root of the order of the involved modes (1).

The absorption of the metallic material controls the energy delivered to the target and is wavelenght dependent. Typical curves are given in fig. 1. The absorption depends also on the angle of incidence and on the beam polarization. It is maximum when the beam is polarized parallel to the cutting direction as shown in fig.2 (2).

The role of the injected gas is twofold :

a - It exerts a mechanical action on the metal removing by friction the liquid layer formed by the laser beam. This role is not so obvious and a peculiar dependance on gas pressure can be observed due to the formation of stationary pressure waves between the nozzle and the workpiece. The real pressure at the target is the relevant parameter for cutting as reported in ref.(3).

b - The injected gas can also start an exothermic reaction providing an added heat source to the workpiece. This allows an oxygen assisted laser cut with enhanced cutting speed and deeper penetration. The behaviour of oxy-laser cutting process is more complex. In general one can observe in the cut section a residual roughness with a set of striations having a regular spacing (4). In addition there is a difference in the slope of striations between the top and the bottom layer. An explanation was proposed by using a two sources model (5). The first layer is apparently melt mainly by direct absorption of laser radiation and the cutting speed is polarization dependent up to a certain thickness value. In the second layer the exothermic reaction has an increasing role as proved by the larger roughness and its lack of dependence on laser beam polarization. An analytical description of the process taking into account both the energy and the mass balance was given recently in ref. (6) apparently with good agreement with the experimental data. This approach is still being improved.

Typical values of cutting speed vs thickness are reported for various metallic materials in fig.3 and in table 1. Iron plates with a thickness up to 30 mm. have been cut using a 4 kW CO2 laser (7).

Pulsed beams have usually peak powers several times larger then the average and this allows an increased absorption due to the generation of the "key-hole" regime also with moderate average power. Pulsed cutting is also very effective in practice when the cutting speed has to be slowed down due to complex features of the cutting path. A more controllable cut can be performed by pulsed beams and this technique has a special importance for cutting or scribing ceramics or glasses. The thermal diffusivity of these material is low enough to give rise to problems of crack formation. This problem again can be partially solved with a gradual increase of pulse amplitude.

DRILLING WITH PULSED BEAMS

The process is substantially different from the previous one as each laser pulse melts and vapourizes the material.The ratio of liquid to vapour contained at the walls of the drill cavity depends upon laser intensity, pulse duration and metal properties. This ratio (usually ranging from 3:7 to 4:1) increases with thermal diffusivity and increased difference between melting and vapourization temperature. Experience has indicated that the maximum aspect ratio (ratio of hole dept to diameter) achievable with a single laser pulse has a value around 5. Various defects such as uneven contour, wall cave-in, splatter and slag are encountered as higher ratios are attempted. Multiple-pulse tecniques can be used to achieve aspect ratios up to 10 with good hole characteristics .An advantage of multiple pulses over single pulse drilling is that a series of higher peak power pulses can produce a higher vapour/liquid ratio then a single high-energy pulse yielding the same hole depth. Therefore, multiple-pulse drilling can provide more precise control over hole contour and straightness with the added advantage of achieving greater aspect ratios, if a sufficient number of pulses is used. For instance to generate round holes with good geometry the beam can be rotated between pulses on its axis or slightly off-axis to average out the irregularities.The principal disadvantage is the increased processing time per hole. Drilling is usually best accomplished by pulsed CO_2 or Nd-YAG lasers also on a wide variety of nonmetallic materials including wood, paper, ceramics, plastics, gemstone, rubber and semiconductors(8).

Nd-YAG lasers are specially suited for drilling high reflectivity materials such as Aluminum and Copper.

CUTTING SYSTEMS

The more diffused cutting systems are those designed for plane cuts of plates or sheets of metallic or non-metallic materials. Depending on the sizes of the workpiece the two X-Y degrees of freedom can be given to the workpiece or to the laser head. An extra degree of freedom is usually required for focusing which is usually automatically accomplished just by contact or non contact sensors.

Three dimensional cutting systems are more complex as they have to handle five degrees of freedom. Actually the optical axis of the focusing head has to be oriented always along the normal to the surface of the workpiece. In addition a sixth axis has to be provided to allow the focusing of the cutting head (9). Usually a circularly polarized beam is used to allow the cutting at the same speed in any direction. More sofisticated solutions have also been developed with the plane polarized beam providing a continuos orientation of the polarization plane along the cutting direction but this requires the control of an extra degree of freedom.

In general laser cutting is competitive up to thicknesses of about 5 mm in metals; beyond this value other methods such as plasma cutting appears more cost effective especially if low roughness in the cut is required. For mass production of components with simple shapes such as round parts the punching process with suitable dies proves to be more convenient than laser cutting. Viceversa for complex shapes or limited production laser cutting is preferable and some cutting systems are designed to provide both options.

Laser cutting systems have instead definite advantages whenever batches of limited number of pieces with different shapes are to be manufactured. The laser results are superior also in implementing prototypes of mechanical parts. A computer designed part can be quickly manufactured almost in real time by a laser cutting system which behaves as special output unit of the computer similar to an X-Y plotter.

Laser beam delivery to the cutting head can be accomplished by articulated arms for the polar robots; by a set of movable mirrors for the gantry type robots .

Other kinds of flexible links are under investigation. Such as metallic wave guides and optical power fibers for Nd-YAG lasers(10). A completely new solution will come out in a short time with the hollow robots which are now being developed.

REFERENCES

1) - R.L.Philips et al. Applied Optics vol.22 N.5 ,pg.643 (1983)

2) - F.O.Olsen   Deutscher Verband fur Schweisstechnik Vol.63 (1980)

3) - B.A.Ward Proc.ICALEO 84 Boston Publ.by The Laser Inst. of America

4) - Y.Arata et al. Trans. of JWRI Vol.8 N.2 pg 15 (1979)

5) - M.Lepore et al. Proc. ICALEO 83 Vol.30 pg 159 Publ.by The L.I.A.

6) - D.Schuocker Proc. of Gas Flow and Chemical Lasers Symp. Oxford 84

7) - G.Sepold Proc. ICALEO 83 Vol.30 pg 153 Publ.by The L.I.A.

8) - C.M.Banas et al Proc. of the IEEE Vol.70 pg 556 June 82

9) - A.Delle Piane Proc. LIM - 2 Birmingham March 85 pg 219 Publ. by IFS

10) - M.Jones Proc.of ICALEO 83 Vol.38 pg 149 Publ. by The L.I.A.

11) - D.M.Roessler et al. Appl. Opt. 17,992 , 1978

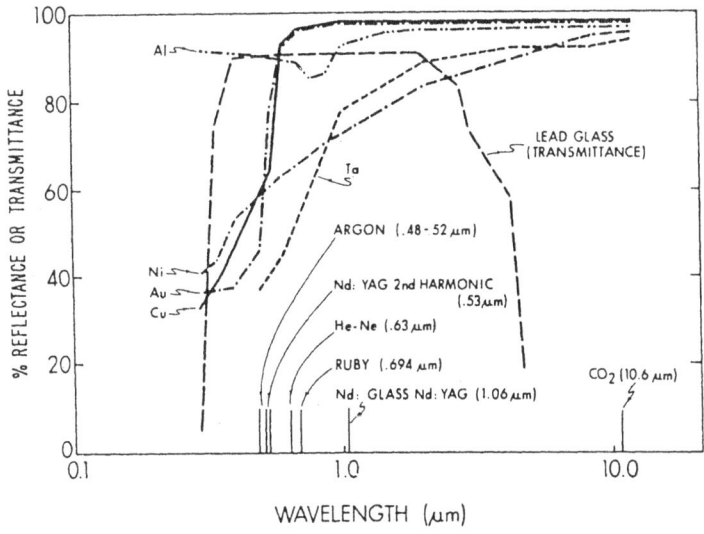

Fig.1 Metallic spectral reflectance

Room-Temperature Absorbtances of Aerospace
Metals and Alloys at 10.6 μm for Various Surface
Conditions and at Normal Incidence

Metal or Alloy	Surface Condition			
	Ideal	Polished	As-Received	Sandblasted
Al	0.013	0.030	0.04 ±0.02	0.115 ±0.015
Au	0.006	0.01	0.02	0.14
Cu	0.011	0.016		0.06
Ag	0.005	0.011		
2024 Al		0.033	0.07 ±0.02	0.25
304 Stainless steel		0.11	0.4 ±0.2	
Ti Alloy (6Al, 4V)			0.65 ±0.2	
Mg Alloy Az-31B			0.06 ±0.03	

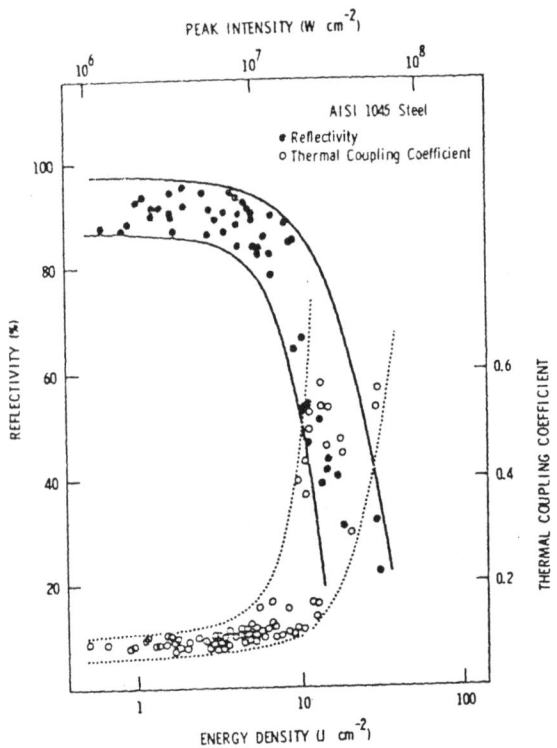

Fig.1b   Reflectivity & thermal coupling for
         AISI 1045 steel at 10.6 microns

(from D.M.Roessler and V.G.Gregson see ref. 11)

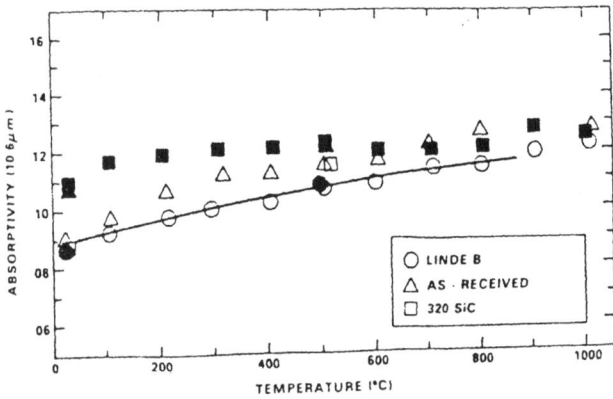

Fig. 1c Absorptivity of 304 stainless steel
        at 10.6 microns vs. temperature

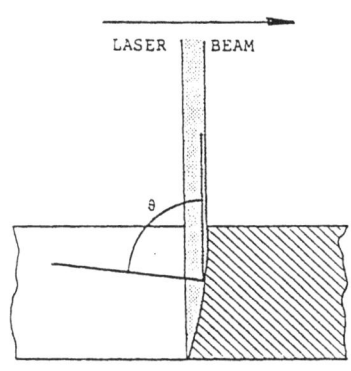

Fig. 2a Incident angle   when
         cutting with laser

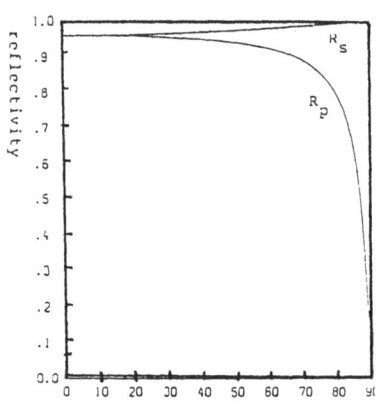

Fig. 2b Reflectivity dependence
        on   for polarized beams

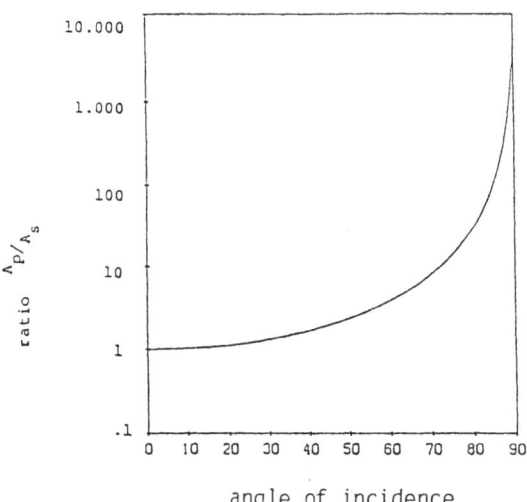

angle of incidence

Fig.2c Absorption ratio versus incident angle of iron
       irradiated by a CO2 laser beam .
       (from F.O.Olsen   see also ref.2)

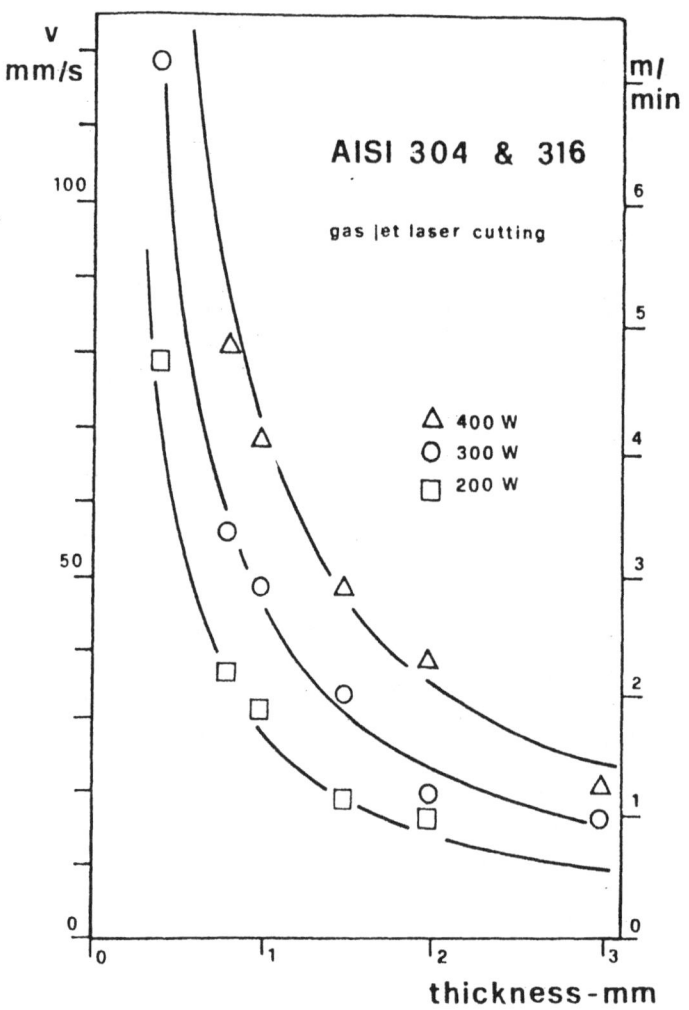

Fig.3 Cutting speed versus thickness for two
stainless steel AISI 304 & 316

$T A B L E\ I$ JET-ASSISTED LASER CUTTING PERFORMANCE

MATERIAL	THICKNESS, in.	CUTTING SPEED, ipm	POWER, kW	SPECIFIC CUT ENERGY, J/in.^{2}
ALUMINUM	0.125	100	4.3	20,600
	0.250	40	3.8	22,800
	0.500	30	6.7	34,200
	0.500	90	15.0	20,000
NICKEL ALLOY	0.125	120	4.0	12,000
	0.500	50	12.0	28,800
STAINLESS STEEL	0.012	173	0.35	10,100
	0.090	70	0.6	9,700
	0.125	100	3.0	14,400
	0.187	50	20.0	128,000
STEEL	0.090	70	0.6	5,720
	0.125	160	4.0	12,000
	0.250	90	15.0	40,000
	0.660	45	4.0	7,750
	2.125	13	6.0	13,000
TITANIUM	0.040	193	0.23	1,920
	0.200	130	0.6	1,390
	1.250	50	3.0	2,880
	2.000	20	3.0	4,500

HEAT FLOW AND FLUID ASPECTS OF CO_2 LASER WELDING

R.C. CRAFER
The Welding Institute, Abington, Cambridge, UK

LASER OPERATION – HEAT AND FLUID FLOW

For the purpose of this analysis, lasers can be regarded as power transformers. Power is supplied in various forms: electrical, chemical, thermal, optical; and light is produced with heat as the main by-product. With few exceptions, the transformation process is very inefficient, so that most of the power is converted to heat. Since most lasers have some limiting values of temperature (or thermal gradient) above which they cease to function, removal of this unwanted heat is critical. In the design of industrial lasers, the useful power output depends on this ability to remove heat.

CONDUCTION LIMITED COOLING – cylindrical geometry

In lasers made from solid materials (e.g. Ruby, Nd^{3+}:YAG, Nd^{3+}:Glass), heat flows through the medium by conduction. Certain low power CO_2 lasers can also be analysed on a conduction-like basis. In these cases, fluid velocities due to natural or forced convection are slow compared to thermal transport velocities due to diffusion.

Consider a laser as a cylinder of solid or gaseous material of radius a and length L, cooled by an external coolant at temperature T_c, Fig. 1.

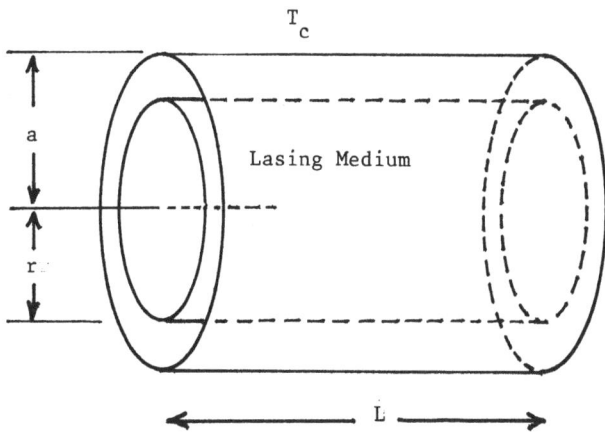

Fig. 1: Cooling model of Laser active medium

Soares, O.D.D., Perez-Amor, M. (eds), Applied Laser Tooling. ISBN-13: 978-94-010-8096-5
© *1987. Martinus Nijhoff Publishers, Dordrecht.*

Assume unwanted heat is deposited uniformly throughout the cylinder at a rate Q per unit volume. Applying Fourier's Law to an axial cylinder of radius r ⩽ a

Heat generated within cylinder = $\pi r^2 Q$ per unit length

Heat loss across boundary at r = $-K2\pi r\frac{\partial T}{\partial r}$ per unit length

where T is the temperature and K the thermal conductivity.

In a continuous power laser these two heat rates must balance, producing the well known differential equation

$$\frac{\partial T}{\partial r} = -\frac{Qr}{2K}$$ _____ [1]

The appropriate solution in this case is

$$T = \frac{Qa^2}{4K}\left(1 - \frac{r^2}{a^2}\right) + T_c$$ _____ [2]

The temperature T is highest when r = 0, that is, along the axis, which is reasonable since this is furthest from the coolant. The axial temperature is

$$T = \frac{Qa^2}{4K} + T_c$$ _____ [3]

Equating this to the limiting temperature T_L and replacing Q by

$$Q = \frac{p}{\pi \xi a^2 L}$$ _____ [4]

where p is the laser power, and ξ the transformation efficiency assumed ⪡ 1, we obtain

$$p = 4\pi \xi KL\ (T_L - T_c)$$ _____ [5]

This is a very important result. It states that the only dimension affecting laser power is length, and that diameter does not enter into the equation. Thus high continuous powers can only be achieved with very long lasers, which is generally not desirable. Of course if the laser is rapidly pulsed, then equation [5] refers to average values, and in that case the power within each pulse can greatly exceed the continuous value for a short time.

CONDUCTION LIMITED COOLING - disc geometry

It is possible to improve matters by building the laser as a series of thin discs of thickness W and area A thus increasing cooled surface area with respect to laser volume, Fig. 2.

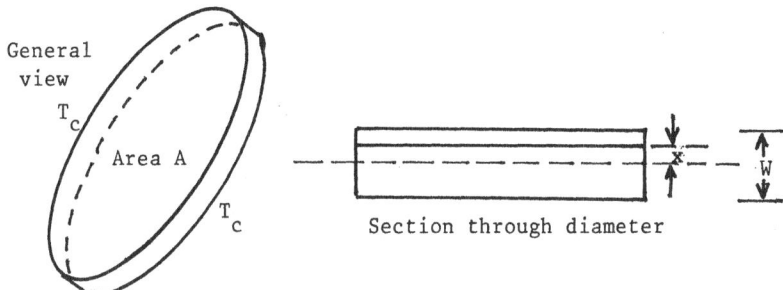

Fig. 2: Disc geometry to increase cooled surface area per laser volume

This is only applicable to solid lasers, and the author is not aware of any instances where such a method has been used for the generation of continuous power. However, in such a case, the differential equation becomes

$$\frac{\partial T}{\partial x} = -\frac{Qx}{K} \qquad\qquad \text{------------------ [6]}$$

where x is the distance from the mid plane of the disc. The final power equation becomes

$$p = \frac{8\xi AK}{W}(T_L - T_c)$$

where A is the area of one face of the disk, assumed cooled on both faces. In this case continuous or average power could be improved by increasing the area A or reducing the thickness W.

CONVECTION LIMITED COOLING - fluid lasers

To achieve really high powers, it is necessary to convect the waste heat forcibly from the lasing region. This method is only applicable to fluid laser media. Although this analysis is principally relevant to fast flow CO_2 lasers, it could also be applied to other gas and liquid laser types. Assume a lasing region of length L and cross sectional area A with a uniform heat deposition rate Q. The fluid laser medium, in this case a low pressure gas mixture of CO_2, N_2 and He flows along the length L at velocity V propelled by a pumping system external to the laser region, Fig. 3.

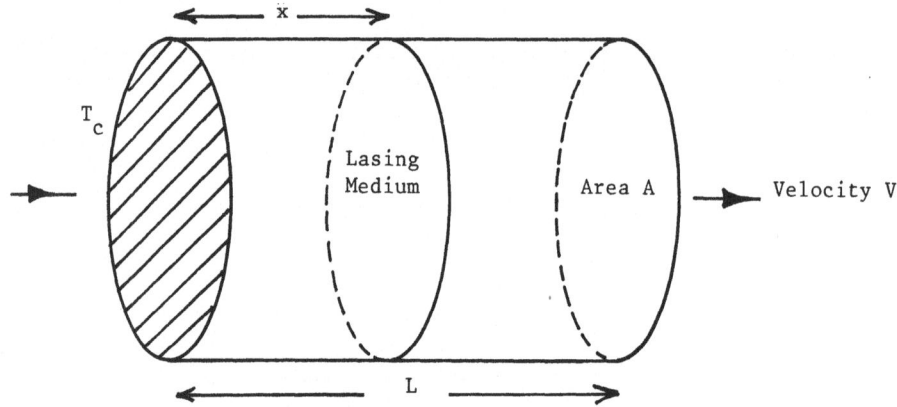

Fig. 3: Convection cooling model geometry

It is assumed that conductive and diffusive heat transfers to the boundaries of the lasing region are negligible in comparison with the forced convention along the length L.

At a distance x parallel to L along the lasing region, the fluid will have acquired an amount of heat per unit volume δH where

$$\delta H = \frac{Qx}{V} \qquad\qquad \text{------------- [8]}$$

This heat will be converted mainly to a rise in fluid temperature δT such that

$$\delta H = \rho c \delta T \qquad\qquad \text{------------- [9]}$$

where ρ, c are the fluid density and specific heat respectively. Equating [8] and [9] and defining the fluid temperature at the entrance to the lasing region as T_c, we obtain

$$T = \frac{Qx}{V\rho c} + T_c \qquad\qquad \text{------------- [10]}$$

The temperature increases linearly along the length of the region, reaching a maximum of

$$\frac{QL}{V\rho c} + T_c$$

at exit. Equating this to the limiting temperature and replacing Q by

$$Q = \frac{p}{\xi AL} \qquad\qquad \text{------------------ [11]}$$

the final power equation becomes

$$p = \xi AV\rho c(T_L - T_c) \qquad\qquad \text{------------------ [12]}$$

In marked distinction to the cylindrical conduction limited case, the power here increases with the cross sectional area and is linearly related to velocity.

DISCUSSION

At the outset it must be stated that all these analyses are only approximate, since they assume constant thermal parameters, constant efficiency and uniform heat deposition. Within these limitations however, the results are surprisingly useful.

For the conduction limited slow flow CO_2 laser, the output power is given by

$$p = 4\pi\xi KL \; (T_L - T_c) \qquad\qquad \text{------------------ [5]}$$

Setting ξ = 0.15 (15% discharge conversion efficiency)
$\quad\quad\quad$ K = 0.14 W/m.°K (Helium at 0°C)
$\quad\quad\quad$ T_L = 250°C (calculated from statistical mechanics)
$\quad\quad\quad$ T_c = 0°C
$\quad\quad\quad$ yields a power per unit length of 66W/m

This compares well with experimental data of 50–100W/m, including the independence on diameter. The calculated figure is based on the limiting temperature being achieved anywhere within a laser cooled only by conduction to the walls. In practice the laser will continue to operate if parts of it are above this temperature, and the laser may produce still higher powers, but the optical quality of the output beam could suffer as a consequence.

For the convection limited laser, the output power is

$$p = \xi AV\rho c(T_L - T_c) \qquad\qquad \text{------------------ [12]}$$

It is perhaps more helpful to rewrite this in the form

$$p = \xi\rho c \; (T_L - T_c) \; \frac{\text{Volume}}{\text{Dwell Time}} \qquad\qquad \text{------------------ [13]}$$

where Volume (= AL) refers to the laser region and dwell time is the time taken for the lasing gas to cross the lasing region.

Setting ξ $= 0.15$
         ρ ~ 0.05 kg/m^3 }
         c ~ 5 kJ/kg   }   approximate values only
         T_c $= 0°C$
         T_L $= 250°C$
         Volume $= 1m^3$
         Dwell time $= 1$ second

we obtain

$$p = 9.4 \text{kW/m}^3/\text{sec}$$

Putting in typical values for a fast axial flow laser, in which each laser tube is about 600mm long, 25mm internal diameter, and the dwell time is about 2msecs, the maximum expected power is about 1.4kW for each discharge. This is consistent with observed values which are typically about half of this predicted maximum.

This form of the power equation also gives an interesting insight into methods of scaling to higher powers. Specific dimensions are no longer important, only volume and dwell time are significant. Thus if technological factors favour a short length and large cross section (cross flow laser), this does not penalise heat removal. In fact in many ways it is an advantage since low pressure fans may be used instead of high pressure blowers as used in axial lasers.

THE WELDING INTERACTION - I HEAT FLOW MODELS

Heat flow within the welding zone is extremely difficult to model accurately. The most successful models to date have utilised techniques such as finite element or finite difference analyses with parameters varying both spatially and temporally. Although such models are capable of considerable accuracy, they are cumbersome, and require extensive computing facilities. If only approximate results are required, then certain simple analytical models may be employed, provided the user is aware of their limitations. The four models (all steady state) to be analysed here are:

1.   Surface heating of thick solid by stationary point source

2.   Surface heating of thick solid by moving point source

3.   Surface heating of thick solid by stationary disc source

4.   In depth heating of solid by moving line/cylinder source.

SURFACE HEATING OF THICK SOLID BY STATIONARY POINT SOURCE

Consider a point source of strength Q watts buried within an infinite solid.

In the steady state, heat crossing any spherical boundary centred on Q must equal Q.

The surface area of a spherical boundary of radius r is $4\pi r^2$, so the heat crossing it, by Fourier's Law, is:

$$- K.4\pi r^2 \frac{\partial T}{\partial r} \qquad\qquad \text{_____} [14]$$

where K and T are the thermal conductivity and temperature respectively.

Equating [14] to Q we obtain

$$Q = -4\pi K r^2 \frac{\partial T}{\partial r} \qquad\qquad \text{_____} [15]$$

Separating variables and solving, we obtain

$$T = \frac{Q}{4\pi K r} + T_{amb} \qquad\qquad \text{_____} [16]$$

where T_{amb} is the ambient temperature

For surface heating, the heat Q from the source is dissipated over half the material, therefore the source term in [16] must be doubled, hence

$$T = \frac{Q}{2\pi Kr} + T_{amb} \qquad\qquad \text{------------} \quad [17]$$

These equations are still relevant for molten material provided K is sensibly constant, and the time scale is sufficiently short to rule out convective heat transfer, that is the molten zone essentially behaves as a solid with respect to heat transfer.

SURFACE HEATING OF THICK SOLID BY MOVING POINT SOURCE

There is no simple solution for this case, so we must resort to an extension of the full theory which is described in standard texts on heat flow, e.g. Carslaw & Jaeger (1).

The steady solution is

$$T = \frac{Q}{2\pi Kr} e^{-\frac{U(r-x)}{2\kappa}} + T_{amb} \qquad\qquad \text{------------} \quad [18]$$

Notice how this collapses to the stationary value $\frac{Q}{2\pi Kr}$ given by equation [17] when U, the velocity = 0.

The approximate form of the isotherms is plotted below, Fig. 4.

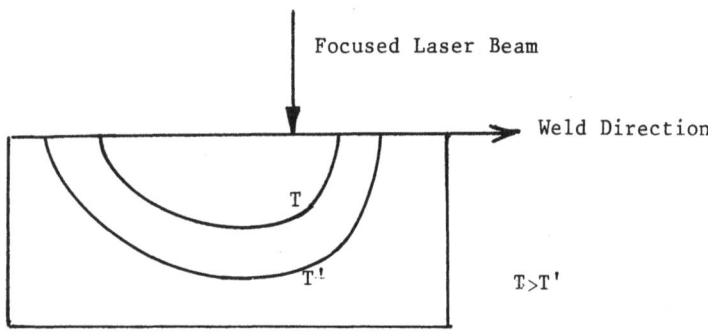

Fig. 4: Isotherms plotted for the case of moving point source

Since these are similar along planes parallel to the source velocity, two things in particular are noticeable:

(a)   The widest part of the thermal profile is behind the source.

(b)   The rate of heating is greatly accelerated in front of the source, and retarded behind.

SURFACE HEATING OF THICK SOLID BY STATIONARY DISC SOURCE

As previously, we perform the calculation for the full infinite solid, then double the power dependent part of the result.

Let absorbed power density at surface be H.

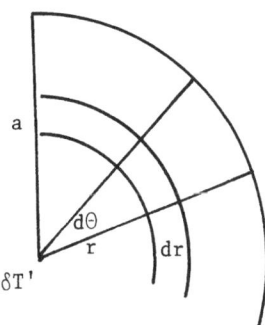

Fig. 5: Disc source geometry used for computation of temperature
distribution

The temperature rise $\delta T'$ at the centre of the disc due to heat conducted from an elementary annulus of radius r is

$$\delta T' = \frac{H d\Theta r dr}{4\pi K r} \qquad\qquad \text{------------ [19]}$$

The full temperature rise T' is

$$T' = \frac{H}{4\pi K} \int_0^{2\pi} d\Theta \int_0^a dr \qquad\qquad \text{------------ [20a]}$$

$$= \frac{Ha}{2K} \qquad\qquad \text{------------ [20b]}$$

Setting the total power $P = \pi a^2 H$ for a uniform disc source
   and $T' = T - T_{amb}$
   and doubling power dependent part for a semi-infinite solid

$$T = \frac{P}{\pi aK} + T_{amb} \qquad \text{------------} \quad [21]$$

Equation [21] enables an estimate of surface temperature for a uniformly heated laser spot.

A similar calculation for a Gaussian beam profile of spot radius a yields

$$T = \frac{2P}{\pi^2 aK} + T_{amb} \qquad \text{------------} \quad [22]$$

IN-DEPTH HEATING OF SOLID BY MOVING LINE/CYLINDER SOURCE

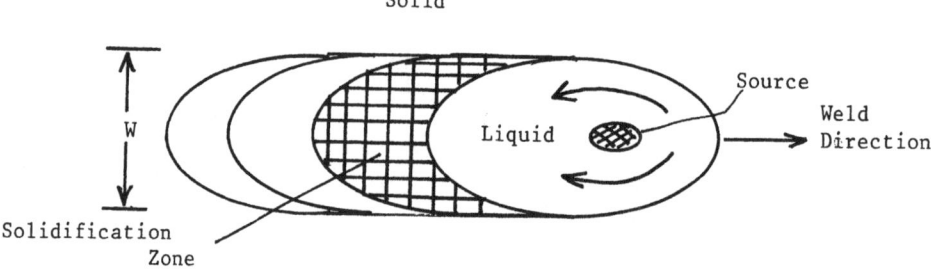

Fig. 6: Plan view of "slice" through weld, thickness δz

This model differs from the previous three in that the dominant heat loss mechanism is assumed to be not conduction, but convection by hot liquid metal out of the weld zone. The line or cylinder source of strength Q per unit length transfers heat to the molten zone and, by its effects, ensures a flow of molten metal from right to left in the diagram. This liquid flow removes heat from the weld zone at a rate

$$V\rho cT_{av} \delta zW$$

Equating this to the heat input rate

$$Q\delta z$$

we obtain

$$W = \frac{Q}{V\rho cT_{av}} \qquad \text{------------} \quad [23]$$

where   W = width of weld
         V = welding velocity
    and T_{av} = average temperature of molten metal.

DISCUSSION

All four analytical models serve to illustrate one aspect of laser welding. The stationary point source model, equation [17], illustrates the thermal distribution within a CO_2 laser spot weld at distances from the surface $\gg$ the size of the focused laser spot. This restriction results from the spot temperature tending to infinity as r tends to zero. Perhaps a more relevant case is the moving point source, equation [18]. This shows the effect of relative motion of source and workpiece, and demonstrates the distortion of the isotherms. At higher powers, the surface vaporises leading to keyhole formation. The critical power and focused spot size required for this can be calculated from equations [21] or [22] depending on which distribution is more suitable.

Having generated a keyhole, the temperature distribution changes from a hemispherical shape to a narrow cylinder shape. The reduced surface area and shorter thermal cycle time mean that conduction ceases to be the dominant process, and is replaced by convection, giving the keyhole equation [23].

Examples

Focused Laser Beam

(1)

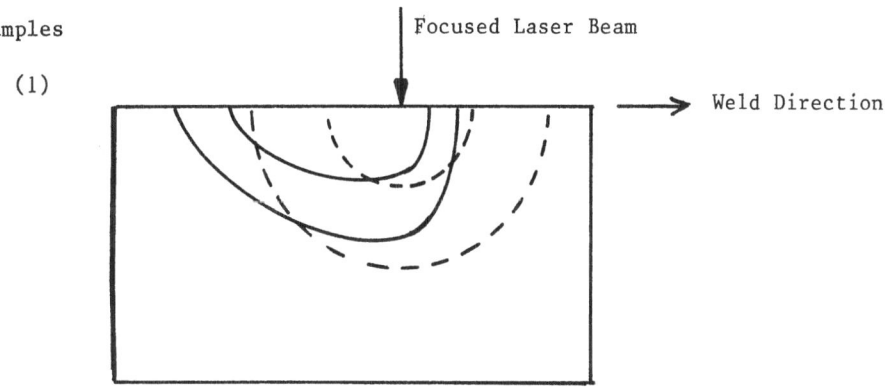

Weld Direction

Fig. 7: Stationary ---------- and moving _____
isotherms for surface heating

(2) Power required to vaporise surface for keyhole initiation. CO_2 laser,

focused spot radius = 150μm
boiling temperature = 2750°C
K ∼ 30W/mK^{-1} (= value just below melting point)
Predicted Power ∼ 400W (assumes 10% absorption)

(3) Width of molten deep penetration weld. (C-Mn steel)

$$Q = 6 \times 10^5 W/m$$
$$V = 1.5 \times 10^{-2} m/s$$
$$\rho = 8 \times 10^3 Kg/m^3$$
$$c = 6 \times 10^2 J/kg/K$$
$$T_{av} \sim 2 \times 10^3 K$$

Gives W ∼ 3mm

THE WELDING INTERACTION - II BEAM ABSORPTION AND FLUID FLOW

Having discussed some heat flow processes both within the laser and the weld, this section describes the physical bases of laser beam absorption both at surfaces and within hot gaseous media, and present possible machanisms for fluid flow within the weld zone.

Maxwell's Equations

The propagation of electromagnetic waves is based on 4 equations:

Gauss' theorem applied to electrostatics $\quad$ div $\vec{D}$ = ρ $\qquad$ ---- [24]

Gauss' theorem applied to magnetostatics $\quad$ div $\vec{B}$ = 0 $\qquad$ ---- [25]

Faraday's and Lenz's Laws of induction $\quad$ curl $\vec{E}$ = $-\dfrac{\partial \vec{B}}{\partial t}$ $\qquad$ ---- [26]

Ampere's Law modified by Maxwell $\quad$ curl $\vec{H}$ = $\sigma \vec{E} + \dfrac{\partial \vec{D}}{\partial t}$ $\qquad$ ---- [27]

Taken together these are known as Maxwells's Equations.

Absorption at a surface

Equation [27] can be rewritten as

$$\text{curl } \vec{H} = \varepsilon \varepsilon_o \left(\frac{\partial \vec{E}}{\partial t}\right) + \sigma \vec{E} \qquad\qquad \text{[28]}$$

where all the symbols have their usual meaning.

The first term on the right hand side of [28] $\left(\varepsilon \varepsilon_o \dfrac{\partial \vec{E}}{\partial t}\right)$ represents 'displacement current' and describes propagation in insulators such as windows, lenses and prisms. Propagation by this means alone is lossless.

The second term $(\sigma \vec{E})$ represents 'conduction current' and describes propagation in conductors and metals, and reflection at mirror surfaces. Propagation by this means is lossy.

For a laser operating at a single frequency f the above equations reduce to:

$$\frac{n}{c} E_y = \mu \mu_o H_z \qquad\qquad \text{[29]}$$

$$\frac{n}{c} H_z = (2\pi f \varepsilon \varepsilon_o - j\sigma) E_y \qquad\qquad \text{[30]}$$

If the laser frequency f $\ll \dfrac{\sigma}{2\pi \varepsilon \varepsilon_o}$ then the conduction current term in equation [30] dominates and the beam is heavily absorbed. (For a typical metal, f $\sim 10^{15}$Hz and for the CO_2 laser, f $\sim 3 \cdot 10^{13}$Hz).

$$\text{The solution is } E = A \exp\left(-\frac{x}{\delta}\right) \exp\left[2\pi f j\left(t - \frac{x}{\delta}\right)\right] \qquad \text{[31a]}$$

where $\delta = (\pi\sigma f\mu\mu_o)^{-\frac{1}{2}}$ is called the skin depth. _____ [31b]

The electric field amplitude diminishes over a distance δ and therefore most laser heating occurs within a small multiple of this distance.

Examples

Typical metal $\sigma = 10^7$, CO_2 laser $f \sim 3 \cdot 10^{13}$Hz
skin depth $\sim$ 30nm

Typical metal $\sigma = 10^7$, YAG laser $f \sim 3 \cdot 10^{14}$Hz
skin depth $\sim$ 10nm

The absorbed energy is thus deposited within the first few atomic layers of the material surface.

A full solution shows that energy not absorbed is reflected with a coefficient

$$R = 1 - \frac{4}{\mu_o\sigma c\delta}$$ _____ [32]

using the above value for a CO_2 laser, and a clean, cold metal surface

$R \sim 96\%$

Warning – in practice R will be lower due to electronic transitions, other absorption processes, surface irregularities and contamination, and will also fall as σ decreases with temperature.

Absorption in a hot gaseous medium

From an analysis similar to the above, it can be shown that hot metal vapours (plasmas) also exhibit a critical frequency called the plasma frequency given by

$$f = (\frac{ne^2}{4\pi\varepsilon_o m})^{\frac{1}{2}}$$ _____ [33]

$\approx 9.0n^{\frac{1}{2}}$

where n is the electron density.

For laser beams below this frequency, the plasma reacts rapidly to electric fields, and the beam is largely reflected. Above this frequency the beam can propagate through the medium and be absorbed over a large distance by various collision processes both within the plasma and its surroundings.

EVOLUTION OF THE WELD ZONE

Stage 1

The intensely focused laser beam ($\sim 10^4$W/mm^2) impinges on the metal surface. Most of the beam is initially reflected, equation [32], but part is absorbed in the top few atomic layers, equation [31b]. Heat is conducted away into the bulk of the metal setting up a time dependent hemispherical temperature profile in the solid. As the metal heats up, its resistivity increases, and hence its absorptivity, as defined by equation [32], increases.

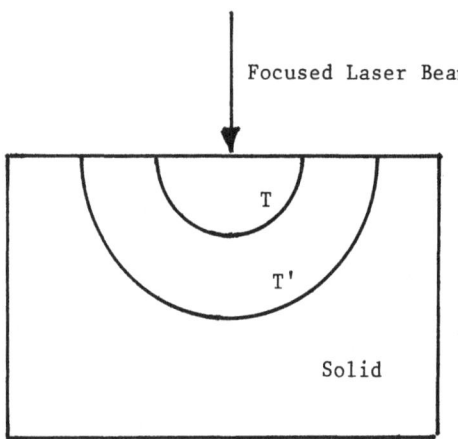

Fig. 8: Time dependent hemispherical temperature profile induced by focused laser beam

Stage 2

For low powers, or defocused beams, the metal will melt but not vaporise, giving rise to a roughly hemispherical molten zone, equation [17]. In practice, convection within the weld pool and other effects will distort the shape away from spherical. This type of weld is known as 'conduction limited' and is characterised by a width approximately twice the depth. Similar shape welds are made by many arc welding processes.

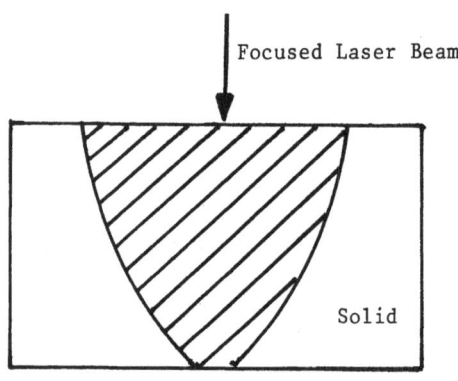

Molten Metal

Fig. 9: Conduction limited type of weld

Stage 3

At higher powers and near to focus the metal surface may boil, equation [21] or [22] (depending on thermal and electrical parameters) and three related events take place.

1. Vapour is liberated which depresses the liquid surface due to reaction pressure. The depressed surface improves beam absorption, by forming an absorbing cavity.

2. The vapour becomes ionised and further beam absorption takes place directly through 'Inverse Bremsstrahlung' (photon-free electron collisions), and indirectly via absorption in the liquid metal surrounding the depression.

3. The depression penetrates completely through the material forming a complete penetration keyhole.

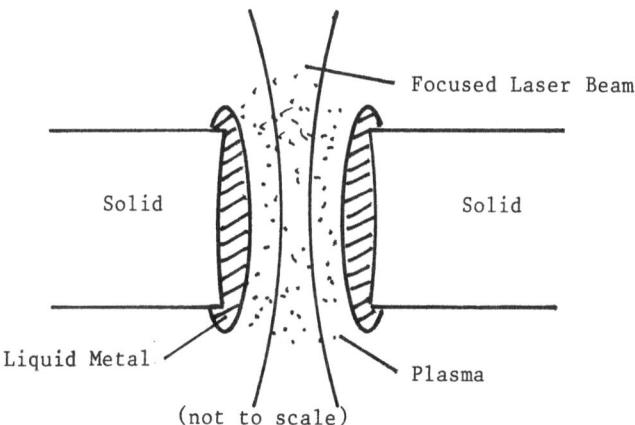

Fig. 10: Penetration keyhole

Motion of the weld zone

For conduction limited welds, and the exposed extremities of deep penetration keyholes, a possible mechanism for fluid flow is surface tension imbalance.

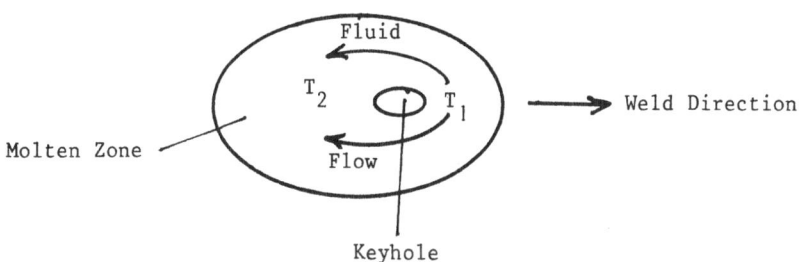

Fig. 11: Liquid metal flow due to surface tension

130

The diagram shows the weld zone, viewed from the direction of the incoming laser beam. At the point of impact, the metal will be at the boiling point. Further away, the temperature will be lower. At temperatures close to boiling, surface tension decreases as the temperature rises. Thus the region marked T_1 will be hotter than the region marked T_2, and there will be a nett surface tension force from T_1 to T_2 with a corresponding liquid metal flow.

Within a deep penetration keyhole, the exposed liquid surfaces are all likely to be at the boiling point, and thus surface tension effects will be minimal. The probable cause of fluid flow here is the difference in evaporation rates. The liquid surface ahead of the laser beam will evaporate at a higher rate than the surface behind the beam, and the difference in reaction pressures will cause fluid flow from front to back of the keyhole, Fig. 12.

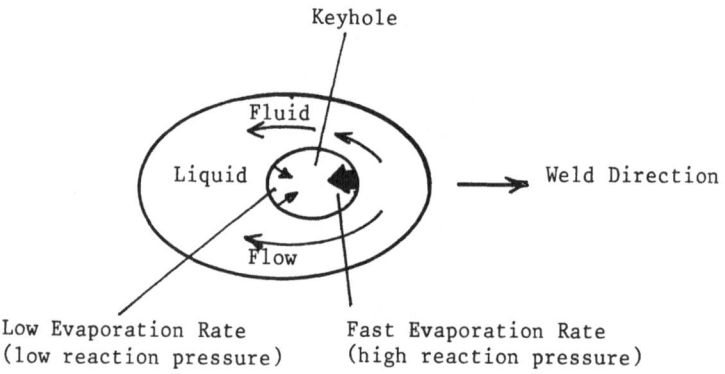

Fig. 12: Fluid flow induce by the difference of evaporation rate

The above explantions for fluid flow are simplistic and highly approximate, but give some physical insight into the mechanisms involved. Readers interested in a more rigorous approach are referred to reference 3.

ACKNOWLEDGEMENTS

The author is indebted to Mr P.J. Oakley and M. Amin for technical discussions, and to Mrs I.J. Harvey and Mrs A.J. Bell for preparation of the typescript.

REFERENCES

1. Conduction of Heat in Solids - H.S. Carslaw and J.C. Jaeger, 2nd ed., OXFORD Press, 1959.
2. Electricity and Magnetism - B.J. Bleaney and B. Bleaney, 2nd ed., OXFORD at the Clarendon Press, 1965
3. The flow of heat and the motion of the weld pool in penetration welding with a laser - J. Dowden, M. Davis and P. Kapadia, Journal of Applied Physics, Vol. 57, pp 4474-4479, 1985.

SURFACE ENGINEERING WITH LASERS

I- Laser Transformation Hardening (LTH)

II- Laser Surface Melting (LSM)

III- Laser Surface Alloying (LSA)

IV- Laser Surface Cladding (LSC)

W.M.Steen,
Metallurgy Department,
Imperial College,
London SW7 2BP

General Introduction

The laser produces a near monochromatic electromagnetic radiation beam usually with a very low divergence angle allowing it to be focused to near theoretical diffraction limited spot sizes. This pure optical energy which is easily focused represents one of the chemically cleanest forms of high energy density available to industry today.

The power from the beam can be controlled rapidly through the drive circuits on the laser, while the power density can be varied from 1 to 10 W/mm by also varying the spot size on the target material. The shape of the beam can be varied by simple optics or moving mirrors. In fact, not only can the laser deliver higher power densities than almost any other energy source available, but it is also one of the most flexible energy sources. This great control over the power density and location allows the laser to be used with surgical precision in surface treatment.

The range of processes the laser is currently used for are illustrated in fig 1. The low power density processes of transformation hardening and Laser Chemical vapor deposition (LCVD) rely on surface heating without surface melting. Processes which rely on surface melting require higher power densities to overcome the conduction heat losses; such processes include simple surface melting to achieve greater homogenization or very rapid self quench processes as in laser glazing for the formation of metallic glasses in certain alloys. The melting processes also include those where a material is added either with a view to mixing into the melt pool as in surface alloying and particle injection or with a view to fusing on a thin surface melt as in cladding. If very short pulses of power of great intensity hit a surface they are able to send mechanical shock waves through the material resulting in surface hardening similar to shot peening but without the surface distortion. This last process of shock hardening has so far been little explored but may eventually find a few exotic processes where it could compete industrially. To date these have not been found. However the other processes are all either in production or being very seriously considered for production.

Range of processes   transformation hardening
                            surface melting
                            surface alloying
                            surface cladding
                            Particle injection
                            LCVD Laser Chemical Vapor Deposition
                            LPVD Laser Physical Vapor Deposition
                            Enhanced Plating
                            shock hardening

    The common advantages of laser surfacing compared to alternative
processes are:

        * Chemical cleanliness.
        * Controlled thermal penetration and therefore distortion.
        * Controlled thermal profile and therefore shape and location of heat affected region.
        * Less after machining, if any is required.
        * Remote non contact processing is usually possible.
        * Relatively easy to automate.

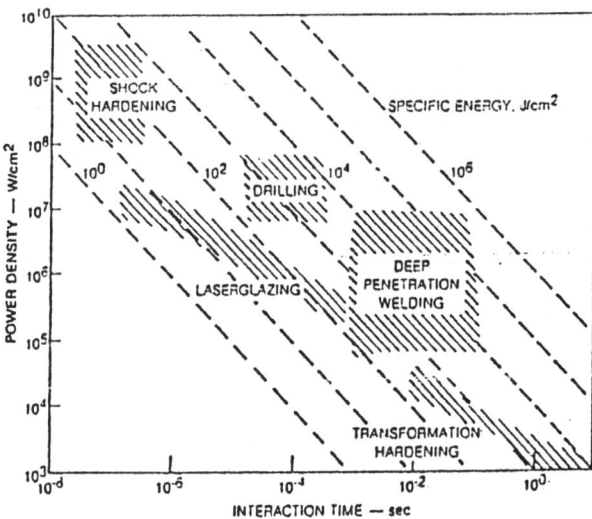

Fig.1: Operational regimes for various laser materials
       processing techniques

I. Laser Transformation Hardening (LTH)

Summary:

    Description of the process
        experimental arrangement
        main characteristics

    Process variations:
        reflectivity control
        optical arrangements, beam spreading
        In-process monitoring

    Current Applications:

    Comparison with other processes:

    Process mechanism:
        Heat flow with no melting
        size of affected zone
            Laser diagrams
            Covering rate vs power vs depth vs hardness
            mass flow

    Product:
        Mechanism of structure formation
        Microstructure
        Properties   hardness
                       fatigue
                       wear

Historical Introduction:

In the early 1960 some work was proceeding on laser transformation hardening (LTH) with a laser in Germany (De Michelis 1970 IEEE J.Quantum Electron. QE-6, 10, 630). The first use in USA was by the US Steel Co. (Speich et al 1966). In 1968 the Russian literature also reported LTH. General Motors Saginaw Steering Gear Division were the first to use LTH in manufacturing around 1970. They were soon followed by Ford who attempted to use the laser to weld auto body components. The initial reason for trying LTH was to harden against wear. The current practical uses include:
1. Hardness increase
2. Strength increase (19)
3. Facilitate lubrification
4. Wear reduction (20)
5. Reharden martensitic Stainless Steel
6. Temper metals
7. Increase Fatigue life
8. Surface carbide creation
9. Creation of unique geometrical patterns

Description of the Process

Experimental arrangement:

Transformation hardening is achieved by passing a laser beam over the surface of a hardenable material - usually a ferritic steel. A thin surface layer is thereby rapidly heated to above the austenitising temperature in the short time that the beam is incident. Once the beam has passed this surface layer is quenched by the conduction of heat into the still cold bulk material of the component. The quench rate is usually fast enough to give hardening by the formation of martensite without the need for an outside coolant such as a water spray (Fig 2). The surface is usually treated to reduce the reflectivity. Gas shrouding is not normally required because there is no melting.

The process is confined to those materials which exhibit some solid phase transformations and whose transformed structures quench to a harder structure than previously. In fact carbon steels and pearlitic cast irons are the usual materials to be treated in this way. Table 1 lists some materials that can be treated and the approximate hardness that would be expected:

Table 1 Materials which can be transformation hardened.

material	transformation temperature C	minimum cooling rate C/s	expected vickers Hardness Hv	melt zone Hv
pearlitic gray cast iron			450-600	700-900
ferritic gray cast iron	800-900		450-550	650-900
S.G.Cast iron			400-500	600-850
Si Al			350-700	850-1020
.1% C Steel			40 Rc	
.2% C Steel			45 Rc	
.3% C Steel (En 8)			50 Rc	
420 Stainless Steel (martensitic)				

The general experimental arrangement is shown in fig 3. (7,32).

There are numerous review articles written on this subject
(6,7,16,24,25,26,29,36,40,41,42,43)

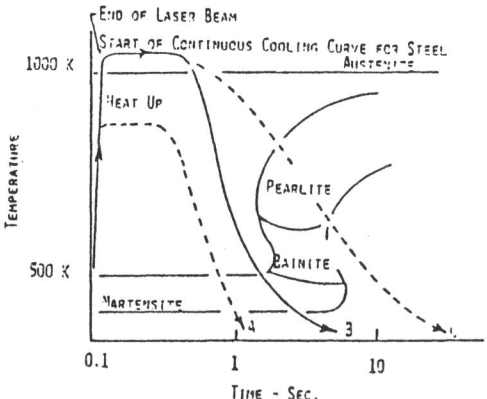

Fig 2 - Metallurgical description of laser beat treatement (52)

TIME - TEMPERATURE CONSTRAINT TO PRODUCE MARTENSITE

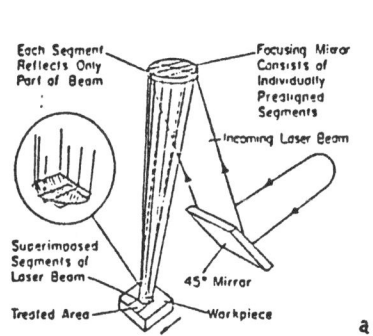

Fig 3a- Beam manipulation technique used for laser, segmented mirror system

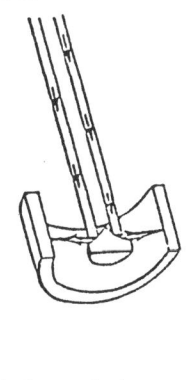

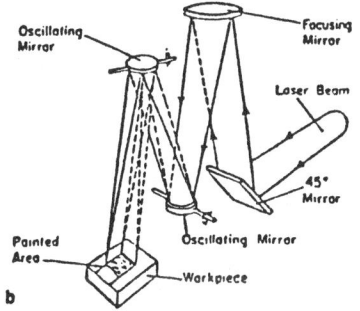

Fig 3b- Beam manipulation technique used for laser, two-axis vibrator system

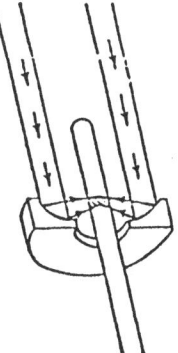

Fig 3c- Toric mirrors for treating a) outer sur-surface of cylinder or b) inner surface of hollow cylinder (7)

Main Characteristics:

Laser heat treatment offers the following advantages: (40,24)

* The treatment can be localized to a required area or pattern
* The heat input is low and confined to a thin surface layer, giving minimal distortion and a reduction in the maximum forces in the component.
* There is no surface disruption, except a slight volume increase due to the phase change.
* No quenchant is required except on very small parts.
* The treatment can be done on the finished article, because distortion is low, surface disruption is nearly zero and there is no contamination from a coolant. (Low temperature stress relief prior to treatment may be necessary with some machined components) (11)
* Accurate control of the treated depth is possible.
* Power density distribution can be controlled allowing the shaping of the hardness depth profile.
* Any area which can be seen, even if only with the aid of mirrors, can be treated.
* The process is easily automated.
* Complex shapes can be treated with only software changes.
* In-process control is a possibility.
* Long beam paths allow optics to be kept at a distance from the processing zone for certain arrangements.
* Process speeds are relatively high, leading to high productivity.
* Rapid quench leads to finer grained microstructures.

Process variations:

1. Reflectivity Control:

Table 2 lists the reflectivity of a number of surfaces.

Table 2
Typical values of reflectivity for various surfaces to 10.6 μm radiation at
normal angles of incidence.

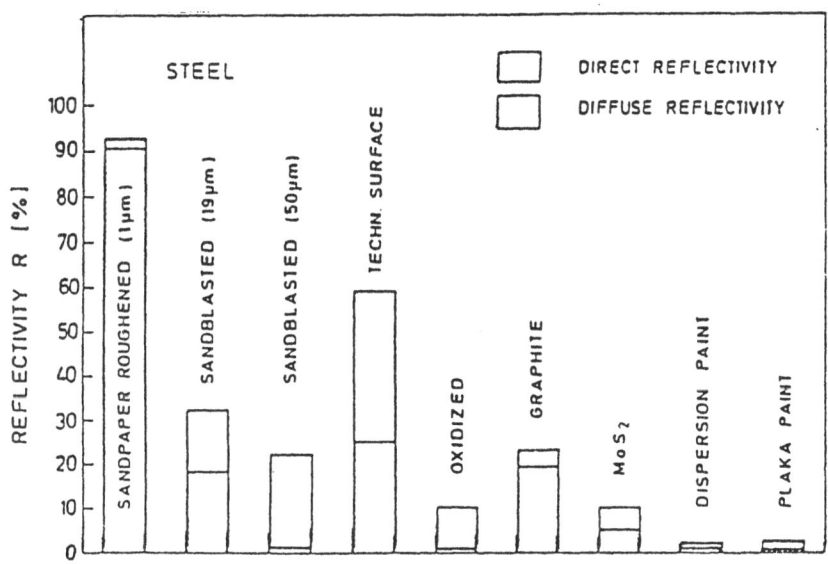

Measurement of direct and diffuse reflection (47)

It will be seen that machined surfaces have reflectivities usually
over 85% and for steels it may rise as high as 95%.

This not only causes a serious loss of efficiency but the reflected
radiation is still concentrated enough to damage the focusing optics or
other equipment if not properly directed. A reduction in reflectivity is
usually achieved by painting the surface with some absorbing agent. The
reflectivity of some typical coatings are also given in table 2 and
illustrated in fig 4 (24). Some coatings have even been patented (21). The
coating operation may be another unit process in a production line in which
case there is a cost penalty as well as the added nuisance of a chemical on
the surface which may need cleaning. Most production routes, however do not
need a special process; for example the material may come from the foundry
already coated as a result of the casting process as for the manganese
phosphate coating in the Saginaw steering gear application.

The coating, if applied separately, is usually sprayed on or painted
on.

There are three ways which have so far been devised to avoid the need
for coating. They are to:

* Shot blast the surface - but this alters the surface finish which
  may be unacceptable.

* Shroud with a reflective dome (36,46).

* Use a polarized beam incident at the Brewster angle (47).

The variation of reflectivity with angle and the plane of polarization is

138

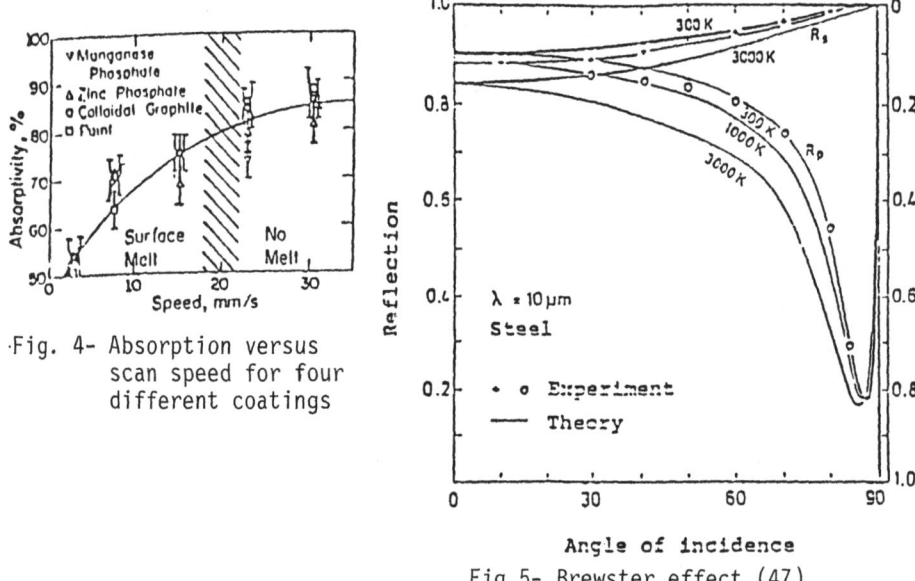

·Fig. 4- Absorption versus
        scan speed for four
        different coatings

Fig 5- Brewster effect (47)

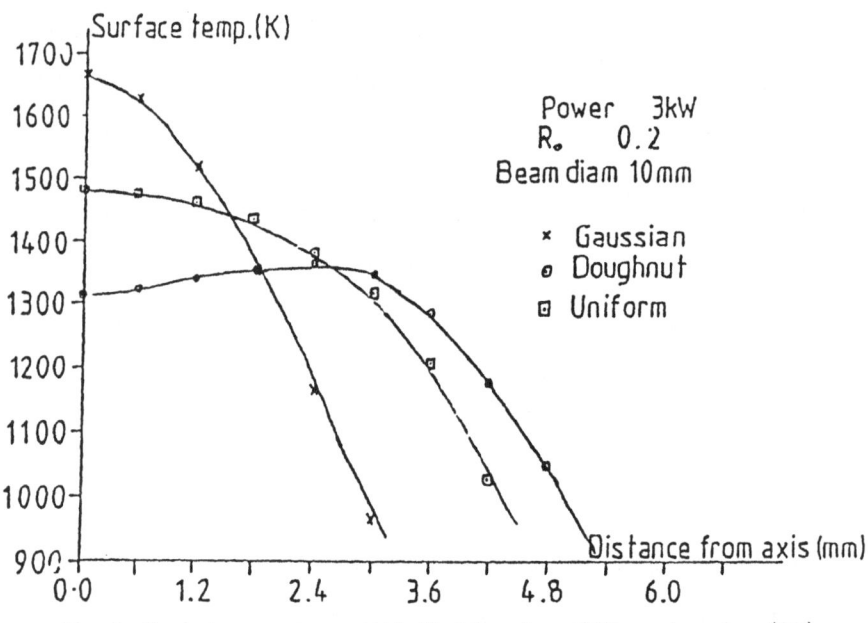

Fig 6- Peak temperature distribution for different modes (53)

shown in fig 5.

2. Optical arrangements:

The process relies upon control of the power density within fairly fine limits if hardening is to occur without surface melting. This is illustrated in the operating curves shown later.

Thus the beam has to be spread over an area and preferentially with a defined power distribution. The heat penetration with different power distributions is illustrated in fig 6. (53,28).

Methods of beam shaping are shown in fig 3:

* Defocusing
* Rastoring in on dimension
* Rastoring in 2 dimensions
* Focusing via an integration mirror
* Focusing via an 'axicon' lens (48)
* Focusing via special optics for special shapes e.g. rods, cylinders
  using 'toric' mirrors.
* Waveguides

This wide variety of methods illustrates the flexibility of optical beam shaping.

3. In-process Monitoring:

The importance of the laser in production will probably be found in highly automated systems. If this is so then methods available for in-process monitoring and therefore possibly in-process control are going to be crucial to the ultimate success of the laser in transformation hardening.

Three methods have so far been explored:
a) Surface temperature measurements:
   This can be done with thermocouples attached to the workpiece or by infra red thermometers. This later process shows considerable promise since it is a non contact method.
b) Acoustic emission measurements:
   It has been shown (49) that the martensitic transformation can be 'heard' via an acoustic emission piezo-electric detector.
c) Measurement of back reflection:

The diagnostic information available from a back reflection signal is extensive and not fully explored as yet. However the acoustic power meter developed recently at Imperial College and now marketed by Quantum Laser allows this signal to be readily measured (50).

Current Applications:

Table 3 Current Applications

Component	Material	Company	Power	Depth	Time/pt	Ref.
crankshaft	nodular cast iron	AVCO		1mm		39
valve guide	gray cast iron	Ford				40
Steering gear housing	ferritic	GM	500W	.5mm	20s	17,33,35
	malleable		1000W			31
	Castings					
Diesel engine liner	Cast iron	GM	5kW			34,39,10
Rivest, clutch						33
Springs						
Bushes						
Camshafts	Cast iron			1.27mm		38
Typewriters Interpolar bars						33
Electric shavers						33
Cut out cams Naval guns	4340 steel	US Navy	1.2kW	.38mm		33
Corrugated paper rolls		Kraft	1.2kW	1mm		33
Engine air intake ports						39
Spline Gear root	1059 steel		500W	.38mm		33,38
Track guides M1-M60 combat vehicles			4kW	.3-.5mm		33
Diesel Engine Piston grooves	.4%C steel S.G.cast iron		5kW	.6mm		32,33,37
Spacers	Cast iron					37
Shafts						38
Cutting edges				.61mm		32
Parking brake bracket						33
Piston rings						2

Comparison with alternative processes:

A summary of the main competing processes to laser transformation hardening - i.e. hardening of the finished article - is given in fig 7. (40,18).

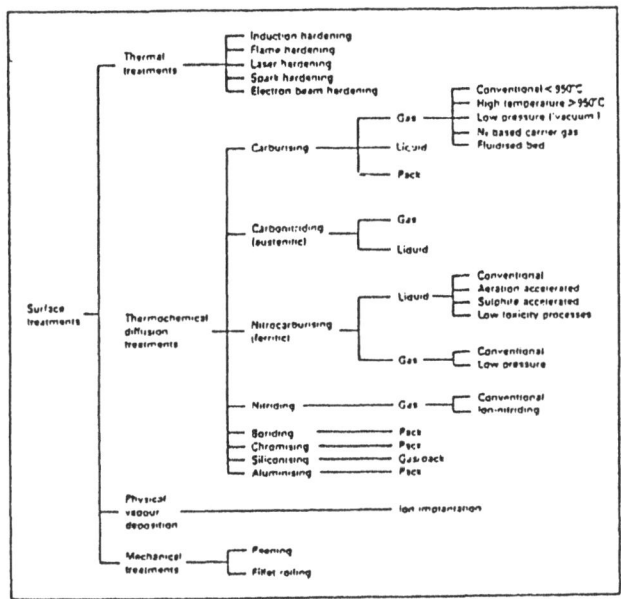

Fig 7- Surface treatments (40)

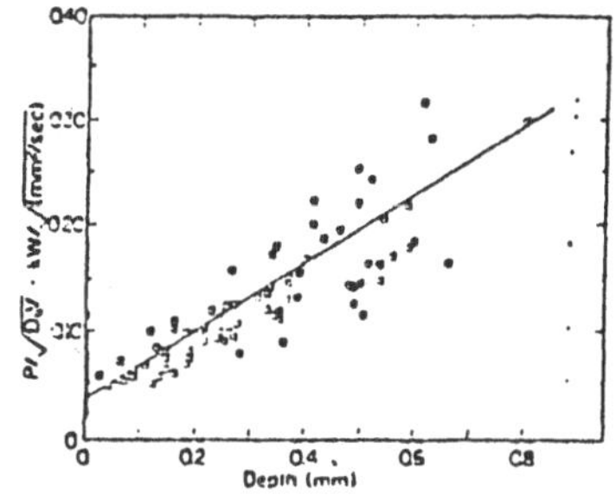

Fig 8- $P/(D \cdot V)^{12}$ versus depth of hardening (27)

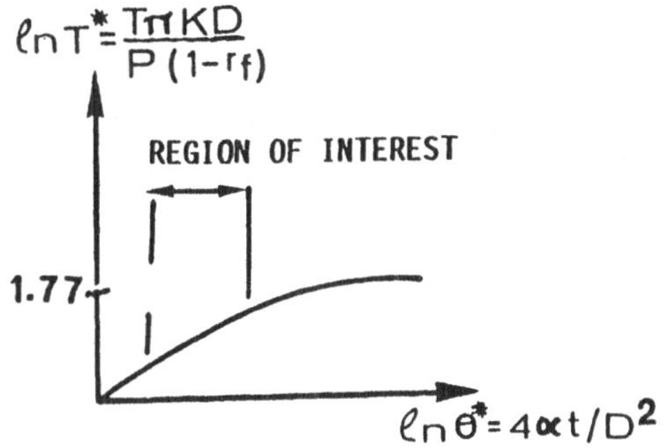

Fig 9- Dimensionless plot of temp: Vs time for a stationary gaus

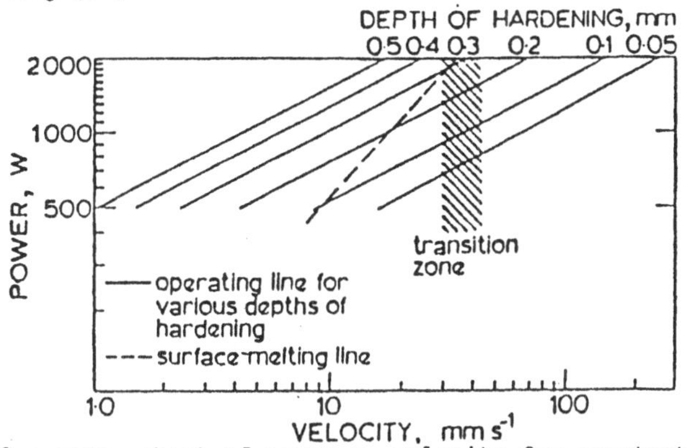

Fig 10- Operating chart of power v. velocity for constant Dn: En 8 steel, constant beam dia 5.8 mm (27)

Process Mechanism:

Heat Flow with no melting:

Transformation hardening with no surface melting is perhaps the simplest laser surface treatment to model mathematically. There is no unknown convection in the melt pool and surface heat losses follow the normal rules of convection and radiation. Numerous models are available (5,16,22,51,52).

Even so it has proved, so far, to be difficult to draw up simple operating diagrams to describe the depth and hardness which are to be expected from a particular treatment. There have been many attempts.

One of the first was the correlation found by Courtney and Steen (27) from a full factorial experimental set in which a correlation between the depth of hardness and the parameter P/DV was found to have a high correlation coefficient fig 8. The reasoning to support this parameter was.

The solution of the temperature under a gaussian surface heat source is shown in fig 9. In the region of the experimental programme noted in the fig it is found that:

$$\ln (T\pi \ kD) \ / \ (P \ (1-r)) = A \ \ln (\ 4\alpha \ D/V \ ) \ / \ d^2)$$

$$\text{from which } T (\quad) = f (P / \sqrt{DV})$$

where α is the thermal diffusivity

This solution is obviously restricted to the small operating region shown. However for En8 steel the relationship found was:

$$d= -0.1097 + 3.02 \ (P/ \sqrt{DV}) \quad \text{mm}$$

where d= depth of hardened zone   mm
       P= absorbed power   W
       D= Beam Diameter   mm
       V= Traverse speed   mm/s

This relationship gives an order of accuracy to the settings on the main operating parameters for any given depth of hardness. The onset of surface melting was found to correspond to a parameter of P/DV. An operating chart based on this data is shown in fig 10.

Many workers have found that the depth of hardening is proportional to the power and inversely proportional to the square root of the velocity but its dependence on the beam diameter is more complex. Sharp et al (53) explored this relationship by means of a finite difference mathematical model. Their theoretical plot of depth vs speed vs beam diameter is shown in fig 11. They note that there are two regimes, one producing cylindrical hardened zones and the other flat sheets of hardness depending upon the beam diameter. In the first case there is two dimensional heat flow while in the second case flow is mainly in one dimension. Thus the search for operating diagrams may be solved by having two sets of curves or a simple mathematical model which would operate on a small home computer or data base (1,12,54).

One solution sought in this later mode is from Ashby and Easterling

(54) and worked out in more detail by Wen Bin Li (55). This solution uses the Rosenthal solution for a moving point source but locates the point source at a fictional height above the true surface in order to simulate a beam diameter and avoid the infinite temperature at the point source itself. Their solutions fit the facts quite well. Their 'Laser diagrams', however plot P/DV vs depth as in fig 12. This is not modeling the effects of either velocity or beam diameter correctly and so a different diagram is needed for each setting of these parameters.

An alternative diagram is that of Gregson (52) fig 13 in which log (P/DV) vs V is plotted from experimental data. Like many such plots it fits the facts but only a small range; with the introduction of higher powered lasers capable of hardening with larger beam diameters extrapolation is not safe.

Mass flow by diffusion:

The transformation hardening of steels the parent structure consists of a non homogeneous distribution of Carbon, in particular. Once austenisation has occurred the carbon moves by diffusion down concentration gradients. The rate of diffusion is described by similar equations to that for heat flow viz:

$$\frac{\partial c}{\partial t} = D_{AB} \left(\frac{\partial^2 c}{\partial x^2} + \frac{\partial^2 c}{\partial y^2} + \frac{\partial^2 c}{\partial z^2} \right)$$

The diffusivity of carbon in austenite is approximately:

$$D = 1 \times 10^{-5} \, e^{-9.0/T}$$

and in ferrite:

$$D = 6 \times 10^{-5} \, e^{-5.3/T}$$

The time for diffusion within the austenite lattice varies with position in the laser treated zone as indicated in the diagram from a theoretical finite difference model fig 14 (27). Thus in laser transformation hardened zones there is always a region around the edges, if not through out, where the carbon has not fully diffused and the resulting structure is a non homogeneous martensite. There have not been sufficient experiments to know whether this non homogeneous martensite is preferable to homogeneous martensite or not. One would lead to higher hardness and therefore better overall wear resistance, a fact which is reported (25) see table later.

Product:

Mechanism of structure formation:

a) Steels: (3,13,14)

144

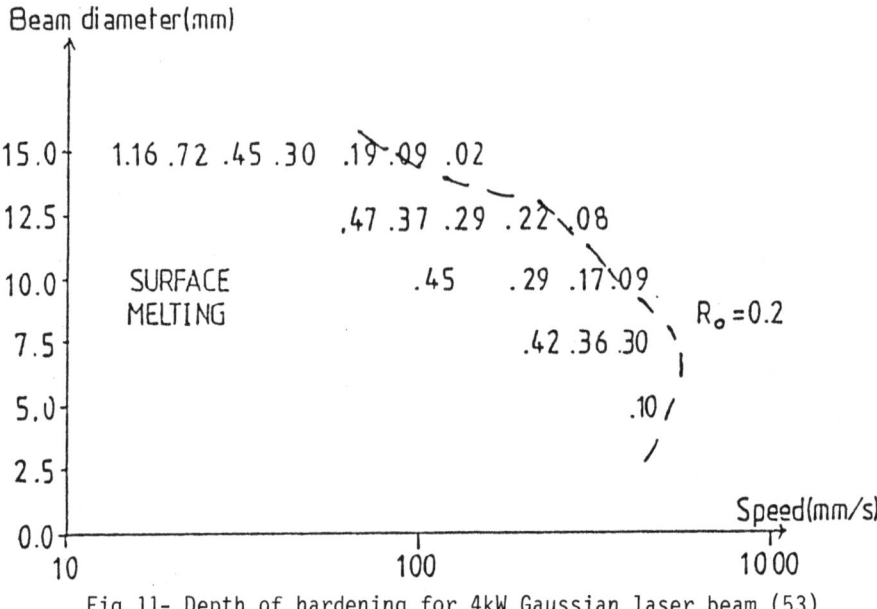

Fig 11- Depth of hardening for 4kW Gaussian laser beam (53)

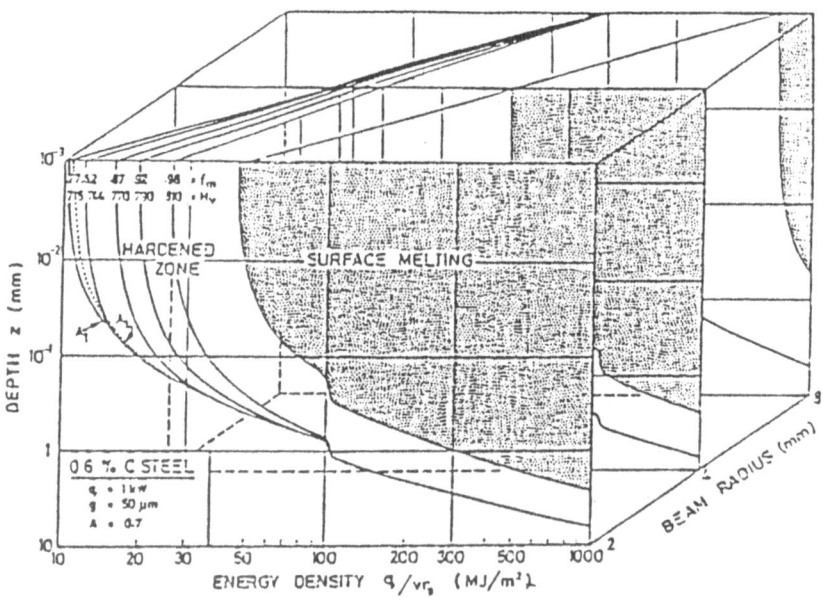

Fig 12- A diagram describing the laser-processing of an 0,6 Wt% carbon
steel. The axes are energy density (g/v r_B), beam radius (r_B)
and depth below the surface (z). (54)

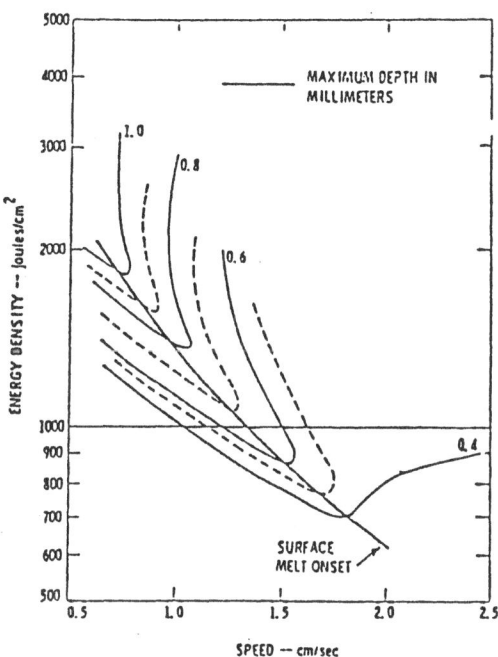

Fig 13- Energy density plot for AISI 1060 steel (52)

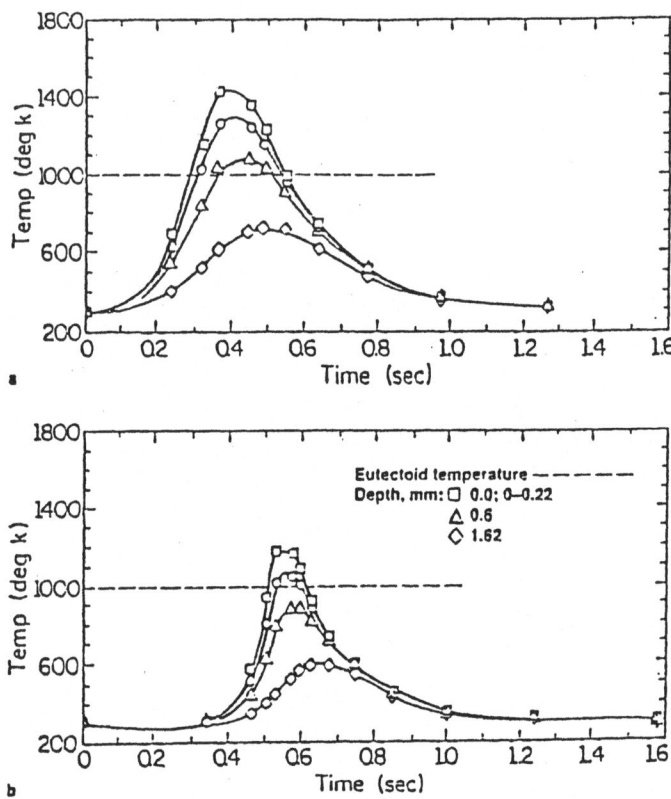

Fig 14- Theoretically predicted thermal cycle during laser heating of En8 (power = 2kW, beam radius = 3.0 mm, and reflectivity = 0.4); a) speed = 22.5 mm/s and b) speed = 42.5 mm/s

On rapid heating, the pearlite colonies first transform to austenite. Then carbon diffuses outward from these transformed zones into the surrounding ferrite increasing the volume fraction of high carbon austenite.

On rapid cooling those regions of austenite having more than a certain amount of carbon (e.g. 0.05%) will quench to martensite, if the cooling rate is sufficiently fast; though retained austenite may be found if the carbon content is above a certain value (>1.0%). The required rate of cooling is indicated in constant cooling curves in which cooling rates from regions in the austenite zone which miss the nose of the nucleation regions for pearlite and bainite would yield martensite, see fig 2. The extent of the conversion to martensite is determined by the martensite start temperature Ms and the Martensite finish temperature Mf.

The transformation of the pearlite is thought to proceed by diffusion from the cementite plates into the ferrite plates, possibly starting from one end of a pearlite colony. The time dependent process does not take long but is sufficient to necessitate some superheat above the austenitising temperature, A1, to allow it to proceed to any extent in laser treatment. The superheat and therefore the extent of the diffusion process is thus slightly affected by the prior size of the pearlite colonies. These colonies, on transformation, become austenite having 0.8% C. The Carbon now diffuses down the concentration gradients into the ferrite regions where there is virtually no carbon. The ferrite regions may also have transformed on the F.c.c. structure of austenite. The extent of homogeneity of the resultant martensite will depend upon the size of the prior ferrite regions and hardness will depend upon the C % as shown in fig 15. (52,23).

It has been stated (41,42) that no unusual metallurgical changes can occur in laser treated steels compared to furnace treated steels. However the more rapid heating and quenching of the laser process does result in variations in the type of Martensite, particularly its fineness, amount of retained austenite and carbide precipitation (43,44,45) as well as the homogeneity of the hardened zone (27).

Cast irons:

Ferritic Gray Cast iron consists of ferrite and graphite regions. As such it is difficult to harden by the laser because the diffusion time is too small. Typically the diffusion distance is 0.1 mm for a 5 mm beam traveling at 20 mm/s. Thus all that is formed is a hard crust around the graphite flakes or nodules. These can still give impressive wear properties though no change in the overall hardness value will be observed.

Pearlitic cast iron is one that has been cooled moderately fast. It consists of pearlite and graphite. In this case transformation hardening is successful in achieving very high hardness levels, as for 0.8% C or higher. With cast irons there is a fairly narrow window between transforming and melting. The irons are important for their ease of casting, that is to say their low melting points; while the A1 temperature is approximately the same for all levels of C.

S.G.Cast iron (8) may result in preferential melting around the graphite nodules due to the lowering of the melting point as the carbon diffuses away from the graphite.

Properties:

Hardness:

This depends upon the C% as shown in fig 15. It has been found that the value may be slightly higher than that found for induction hardening as listed in table 4.

This difference is probably due to the shallower zone in the laser process allowing a faster quench and therefore less homogeneity of the C causing regions of higher C% and therefore harder, as well as an unusually deformed martensite as reported by Kickuchi (25).

Overlapping of successive tracks induces a thermal experience in the neighboring tracks so that there is some back tempering. A hardness trace through such a overlapped region is shown in fig 16. In the example of the Diesel engine cylinder liner this softer region aids in reducing wear it becomes a sink for debris and oil. Patte and hardened surfaces have not received too much attention as wear surfaces mainly because prior to the laser they were difficult to make. The laser can make patterned surfaces easily, and therefore opens up a whole new study in tribology.

Table 4 Experimental Results of Rotational Wear Resistance Tests (25)

	Laser Hardening	Induction Hardening
Material	SK5 (AISI WI)	SK5 (AISI WI)
Hardness	HRC64-67	HRC60-63
Case Depth	0.7-0.9mm	2-3mm
Load	101 kg/mm	101 kg/mm
Scoffing	no occurrence	slight
Wear loss	0.5	1

FATIGUE:

In steels and cast irons there is a residual compressive stress on transformation hardening due to the volume expansion on the formation of martensite. Fatigue cracks are generally initiated at the surface by tensile stresses; thus the fatigue load must be sufficient to overcome this residual compressive stress before a crack can propagate.

Thus several workers have reported improved fatigue life with laser treatment (30,29,25). Gnanamuthu and Singhe measured compressive stresses around 360 to 512 MPa and improved fatigue life a factor of 30% over the untreated material for AISI 1945 steel. Mazumder (7) has fatigue tested rods laser treated without any track overlap using toric mirrors and has found improved fatigue life for gray cast iron in comparison with the same material as cast, or conventionally oil quenched and tempered.

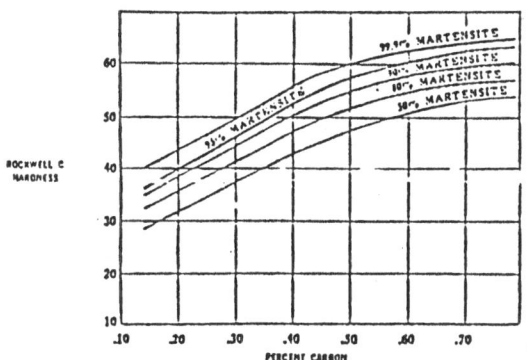

Fig 15- Relationship between carbon content hardness and martensite
percentage (52)

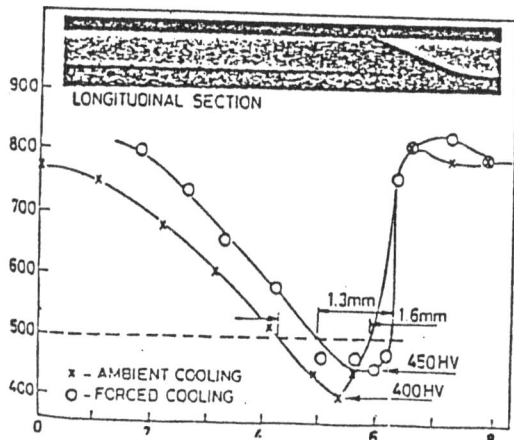

Fig 16- Tempering of previously hardened layer at overlap region in steel
showing beneficial effect of forced air cooling. Hardness measurements taken 0.1 mm below surface. (32)

150

Wear resistance:

The wear resistance of AISI 1045 steel was studied by Gnanamuthu (29) for different heat treatment conditions. He reported a wear loss of 0.6-1.4 mg for laser treated surfaces; 1.35 mg for oil quenched steel (Rc26) and 0.25 mg for water quenched (Rc61). Kickuchi's results (25) on SK5 steel are shown in table 4 where it is seen that the wear resistance of laser treated surfaces was found to be twice as good as that of an induction hardened surface.

References Transformation Hardening:

1. Inoue. K., Matsumura. S., Arate.Y. "Condition Method Utilising Data Base System in CO Laser Surface Hardening", JWRI vol 11 n. 1 june (1982) pp 37-46

2. Ding. C. "Laser Heat Treatment of Piston Rings", Z.Werkstofftech. March (1983) 14 (3) pp 81-85

3. Sepold. G. "Hardening and Alloying of Steel Surfaces by High Powered Laser Beams" Physical Processes in Laser/Material Interactions pp 163-173 Plenum Press New York 1983.

4. Bass. M. "Laser Heating of Solids" pp 77-115 "Physical Processes in Laser/Material Interactions" Plenum Press New York, 1983

5. Kou. S., Sun. D.K. "Heat Flow During the Laser Transformation Hardening of Cylindrical Bodies" Metall Trans A Sept 1983 14A (9) 1859-1867

6. Hick. A.J. "Rapid Surface Heat Treatments - Review of Laser and Electron Beam Hardening." Heat Treat Met. 1983 10 (1) 3-11

7. Mazumder. J. "Laser Heat Treatment: The State of the Art" J. Met., May 1983 35 (5) 18-26

8. Hawkes. I., Steen. W.M., West. D.R.F. "Laser Surface Hardening of Ferritic Sheroidal Graphite Cast Iron" Metallurgia V50 N2 pp 68-73 1983

9. Albert. M. "Laser on a Lathe" Mod. Mach Shop May 1983 55 (12) pp 50-58

10. Andryakhin. V.M., Vasilev. V.A., Sedunov. V.K., Chekanova. N.T. "Effect of Wear Resistance of a Laser Induced Hardening System for Cylinder Liners" (translation) Met Sci Heat Treat (USSR) Sept-Oct 1982 24 (9-19) pp 645-647

11. Goncharenko. V.P., Voronov. I.N., Velikikh. V.S., Kartavtsev. V.S. "Deformation of Steels During Laser Hardening" (translation) Met Sci Heat Treat (USSR) Sept-Oct 24 (9-19) pp 641-645

12. Brativnik. E.V., Velikikh. V.S., Goncharenko. V.P., Kartavtsev. V.S., Tarararksina. O.G. "Method of Determining the Optimum Regimes of Laser Hardening of Steels and of the Quality Control of Hardening" (translation) Met Sci Heat Treat (USSR) Sept-Oct 1982 24 (9-19) pp 637-640.

13. Burakov. V.A., Brover. G.I., Burakova. N.M. "Heat Resistance of Steel R6M5 After Laser Treatment (translation) Met Sci Heat Treat (USSR) Sept-Oct 1982 24 (9-10) pp 634-637

14. Lutsenko. T.I. "Laser Hardening of Tools" (translation) Met Sci Heat Treat (USSR) Sept-Oct 1982 24 (9-10) pp 632-634.

15. Grigoryants. A.G., Safonov. A.G., Tarasenko. V.M., Mareev. N.Y. "Structure and Hardness of Steel 45 After Irradiation With a CO Laser" (translation) Met Sci Heat Treat (USSR) Sept-Oct 1982 24 (9-10).

16. Kou. S., Sun. D.K., Le. Y.P. "A Fundamental Study of Laser Transformation Hardening" Metall. Trans A Apr. 1983 14A (4) pp 643-653.

17. Udall. H.N. "High Frequency Selective Surface Hardening Heat Treat Met. 1982 9 (4) pp 94-95

18. Stahli. G. "Physical and Materials Related problems in Short Time Surface Hardening of Steel Using High Frequency Laser and Electron Beams" Tribol. Int Apr 1981 14 (2) pp 101-106

19. Shur. E.A., Voinov. S.S., Kleshcheva. I.I. "Increasing the Design Strength of Steels by Laser Hardening" (translation) Met Sci Heat Treat (USSR) May-Jun 1982 24 (5-6) pp 341-344.

20. Bell. T. "Surface Heat Treatment of Steel to Combat Wear" Metallurgia March 1982 49 (3) pp 103-111.

21. Lorenzo. R., Wolf. F.J. "Laser Hardening of Steel Workpieces" Patent US 4313771 Feb 1980 .... Surface Coating Alkali Metal Nitrite.

22. Mazumder. J., Steen W.M. J. App. Phys. 51 (3) 1980 pp 941

23. Sadovskiy.V.D., Tabatchikova.T.I., Salokhin. A.V., Malysh. M.M. "Phase and Structure Transformations During Laser Heating of Steel Influence of the Original Structure" Phys Met Mettalurgy V53 n.1 1982 pp 77-84.

24. Trafford. D.H.N., Bell. T., Megaw. J.H.P.C., Bransden. A.S. Heat Treatment(1979) The Metal Society, London (1979) pp 33-38.

25. Kikuchi. M., Hisada. H., Kuroda. Y., Moritsu. K. "The Influence of Laser Heat Treatment Technique on Mechanical Properties" Paper N.12 Proc 1st Joint US/Japan Int Laser Processing Conference Laser Inst. of America Toledo Ohio 1981.

26. Gregson. V. "Laser Heat Treatment" Paper N.15 Proc 1st US/Japan Int Laser Proc. Conf. Laser Inst Am. Toledo Ohio.

27. Steen. W.M., Courtney. C. "Laser Surface Treatment of En8 Steel using 2 KW CW Laser" Metals Tech 6 N.12 456 1979.

28. Sandven. O.A. "Laser Surface Heat Treatment with Profiled Beams" Proc 1st Int US/Japan Laser Proc Conf. Laser Inst Am Toledo Ohio 1981.

29. Gnanamuthu. D.S. "Applications of Lasers in Material Processing" Ed. Metzbower ASM Ohio 1979 pp 202.

30. Singhe. H.B., Copley. S.M., Bass. M. Met Trans A 12A Jan 1981 pp 138

31. General Motors Corp Electromotive Division Lagrange Illinois "Looking in on Lasers" Heat Treating 10 (7) 1978 pp 22.

32. Bransden. A.S., Gazzard. S.T., Inwood. B.C., Megaw. J.H.P.C. preprint CLM-P745, Culham Lab. Oxon. "Laser   Hardening of Ring Grooves in Medium Speed Diesel Engine Pistons" 1985

33. Molian. P.A. "Engineering Applications and Analysis of Hardening Data of Laser Heat Treated Ferrous Alloys" to be published Surface Engineering 1985

34. Strong. E.J. "How General Motors Decided to Heat Treat with Lasers on the Assembly Line" Laser Focus/Electro Optic Nov 1983 pp 172-180

35. Miller. J.E., Wineman. J.A. "Laser Hardening at Saginaw Steering Gear" Metal Progress May 1971 111 38.

36. La Rocca. A.V. "Laser Applications in Manufacturing" Scientific American March 1982 80-87

37. Seaman. F.D., Gnanamuthu. D.S. "Using the Industrial Laser to Surface Harden and Alloy" Metal progress Aug. 1975 pp 210-215

38. Locke. E.V., Gnanamuthu. D.S., Hella. R.A. "Welding and Metal Working with High Powered CO Lasers" Proc. US Metal Proc. and Fab Seminar March 1974 Stratford England p 1-37 ed Robert. M. Silva publ. by Universal Tech Corp Dayton Ohio 45432 USA

39. General Motors Corp "Laser Heat Treating of a Diesel Engine Crankshaft" Source Book on Applications of Lasers in Metal Working ASM Metals Park Ohio ed Metzbower pp 227-228

40. Oakley. P.J. "Laser Heat Treatment and Surfacing Techniques - a Review" Welding Institute Research Bulletin 1981 22 1

41. Sandaven. O.A. "Laser Surface Transformation Hardening" Metals Handbook Vol 4 9th edition, ASM Ohio pp 507-517 1982

42. Engel. S.L. "Basics of Laser Heat Treating" GTE Sylvania Inc. Electro Optics Org. March 1976

43. Gnanamuthu. D.S. "Laser Surface Treatment" Optical Engineering Vol 19 N. 5 (1980) pp 783-792

44. Molian. P.A. "Laser Surface Heat Treatment of AISI 4340 Steel - A Microstructural Study" Materials Science and Engineering Vol 51, N. 21, (1981) pp 253-260

45. Molian. P.A. "On the Presence of Retained Austenite in Laser Transformation Hardened Ultrahigh Strength low Alloy Steels" Scripta Metallurgica, Vol 15, N. 10 (1981) pp 1101-1104

46. Weerasinghe. V.M., Steen. W.M. "Laser Cladding with Pneumatic Powder Delivery" Proc 4th Int Conf on Laser Processing Los Angeles Jan 1983

47. Dausinger. F. "Laser Hardening in Precision Components" Proc Seminar "Exploiting the Laser in Engineering Production" Coventry Sept 1984 publ Welding Institute Cambridge.

48. Oakley. P.J. "Laser Transformation Hardening of a Medium Carbon Steel Pt.2 - Preliminary Assessment of the Axicon Lens" W.I. Res. Report 205/ Feb 1983

49. Rawlings. R.D., Steen. W.M. "Acoustic Emission Monitoring of Surface Hardening by Laser" Optics and Lasers in Engineering Nov. (1981) pp 173-187

50. Steen. W.M., Weerasinghe. V.M. "In Process Laser Beam Characterization" Proc Laser 85 Optoelektronik conf Munich July 1985 paper 1.8.6.

51. Hsu. S.C., Kou. S., Mehrabian. R. "Rapid Melting and Solidification of a Surface Due to a Stationary Heat Flux" Met Trans B Vol. 11B N.1 Mar (1980) pp 29-38.

52. Gregson. V.G. "Laser Heat Treatment" Ch 4 Laser Material Processing ed M. Bass publ. North Holland publ Co. New York 1983

53. Sharp. M., Steen. W.M. "Investigation Process Parameters for Laser Transformation Hardening" Proc 1st Int Conf on "Surface Engineering" paper 31, publ Welding Institute, Cambridge 1985

54. Ashby. M.F., Easterling. K.E. "The Transformation Hardening of Steel Surfaces by Laser Beams" Acta Metall. to be published.

55. Wen Bin Li "Laser Transformation Hardening of Steel Surfaces" Ph.D. Thesis Lulea University, Sweden

II. Laser Surface Melting (LSM)

Summary:

   Description of the process
      Experimental arrangement
      Main characteristics

   Process Variations:
      reflectivity control
      optical arrangement
      shrouding
      applications

   Process mechanisms:
      Solidification Mechanisms
      Heat flow with melting
      G, R predictions
      size of affected zone
      flow within the melt pool
      surface profile
      porosity
      stress cracking and distortion

   Product:
      microstructures cast irons
                      Stainless steel
                      Titanium
                      Steel
      properties   Hardness
                      wear
                      corrosion
                      fatigue

Description of the process:

     The experimental arrangement is similar to that for transformation
hardening or welding. A near focused laser beam is traversed over the
surface to be melted while the surface is shrouded in an inert atmosphere.
Careful control of the power, speed and beam diameter allows control of the
melt pool size and the flow of metal within it.

The Main Characteristics:

   * The melt produced has a fine near homogeneous structure.
     This structure is often harder and more corrosion resistant than the
     untreated surface.

   * Superficial melting and high traverse speeds result in little thermal

penetration. Thus the process can be executed quite close to thermally sensitive   components.

* The highly localized melt zone and low thermal penetration results in minimal thermal distortion

* Surface finish of -25 μm is fairly easily obtained.   Better figures have been recorded particularly with Ni alloys.

* The   process can be executed in difficult locations as noted under transformation hardening.

* There are the usual advantages of the high precision and flexibility of laser power.

Process variations:

1. Reflectivity Control:

Although reflection of the incident radiation is a problem there is less that can be done about it than there was with transformation hardening. The problem is that as melting occurs so the anti-reflection coating is removed. However once the material is hot the reflectivity is reduced as indicated in the Bramson Equation: (43)

$$\varepsilon_\lambda (T) = 0.365 \left(\frac{\rho r \ (T)}{\lambda}\right)^{\frac{1}{2}} - 0.667 \ \frac{\rho r \ (T)}{\lambda} + 0.006 \left(\frac{\rho r \ (T)}{\lambda}\right)^{\frac{3}{2}}$$

where $\rho r \ (T)$ = Electrical resistivity at a temperature T°C
$\varepsilon_\lambda \ (T)$ = Emissivity at T° C
λ = Wavelength of the incident radiation μm

As noted before the reflectivity varies with the angle of incidence (1). Surface films play a significant role. The small addition of oxygen to the shroud gas has a notable effect on the reflectivity as found by Jorgensen 1980 (2). A surface plasma will initially help to couple the beam into the surface. If the plasma leaves the surface then it will block the beam.
Optical feed back systems as described by La Rocca (3) and Weerasinghe (4) can increase the laser coupling by around 40%.

2. Optical Arrangements:

Due to the high reflective energies all metal optics are preferred. These are usually made of Molybdenum since this material is hard and has a high melting point allowing cleaning from the spatter which rises at high speeds from the workpiece. Energy densities of around 10^5 W/cm^2 are required for this process and so near focused beams are required for lasers with powers less than 3Kw. Above this power a linear oscillation of the beam is sometimes employed. The energy density is less than is required for the formation of a keyhole. Hence the laser beam power distribution or mode structure which affects the focusability of the beam is not a crucial factor. Thus lasers having poor mode structures can be used for this process.

Shrouding:

Good shrouding is necessary for process repeatability and good surface finish, apart from the obvious metallurgical aspects.

Several systems have been designed for good shrouding operating in the open atmosphere. They are illustrated in figure 17.

* coaxial jet:
   velocity is limited by the need to avoid surface disruption.
* Chung shroud (5)
   excellent results
* Powell shroud (6)
   fails at high traverse speeds
* Side jets (7)
   excellent results.
* trailing jets.
   always needed for high traverse speeds.

Applications:

1. Surface melting of cast irons:
   Not at present used but the process does produce a very high hardness on a relatively cheap material which is tough in the bulk.
2. Surface melting of Tool steels:
   Again not used at present but a very high hardness is achieved.
3. Laser glazing:
   The formation of metallic glasses is being considered for their good corrosion properties.

The lack of current applications for surface melting is due to two factors:
1. If surface melting is required then surface alloying is almost the same process and offers the possibility of vastly improved hardness, wear or corrosion properties.

2. The very high hardness obtained with cast irons and tool steels are associated with some surface movement and hence require some finishing or machining after treatment. This is not so easy to do with the high hardnesses obtained. Hence the economics of standard heat treatment of the finished shape is sometimes preferable.

Process Mechanisms:

Solidification Mechanism:

Style of solidification: (8)

Solidification will proceed as either a stable planar front or as an unstable front leading to dendrites or cells. Which process occurs depends upon the concurrence of "Constitutional Supercooling" as illustrated in fig. 18. Constitutional supercooling is caused by the thermal gradient being less steep than the melting point gradient, which in turn is the result of the variation of composition at the solidification front due to partition effects taking place there.

Expressed mathematically we can state:

The gradient of the solute in the liquid at the interface is:

$$\left(\frac{d\,C_L}{dx}\right)_{x=0} = -\frac{R}{D_L}\;C_L^{\;*}\,(1-k)$$

The limit before the onset of constitutional supercooling is defined as:

$$\left(\frac{d\,T_L}{dx}\right)_{x=0} = M_L\left(\frac{d\,C_L}{dx}\right)_{x=0} = G$$

where C_L = liquids composition
T_L =   "   Temperature C
R = rate of solidification m/s
D_L = Diffusitivity   m/s
C_L*= Liquids concentration in equilibrium with solids concentration C
K = Partition coefficient
M_L = Slope of the liquids
G = Thermal Gradient   C/m

Combining these two equations we have the general constitutional

supercooling criterion:

$$\frac{G}{R} \geqslant -\frac{M_L\,C_s^{\;*}\,(1-K)}{K\,D_L}$$

Thus the ratio (G/R) should be large for a stable planar front solidification mechanism. Fig. 19 illustrates this equation and further introduces the concept of "Absolute Stability" when the solidification rate R is also large that there is no time for diffusion.

Scale of solidification structure:

If the dendritic or cellular structure is sufficiently fine then it is possible to approximate the liquid between the cells as being a small

158

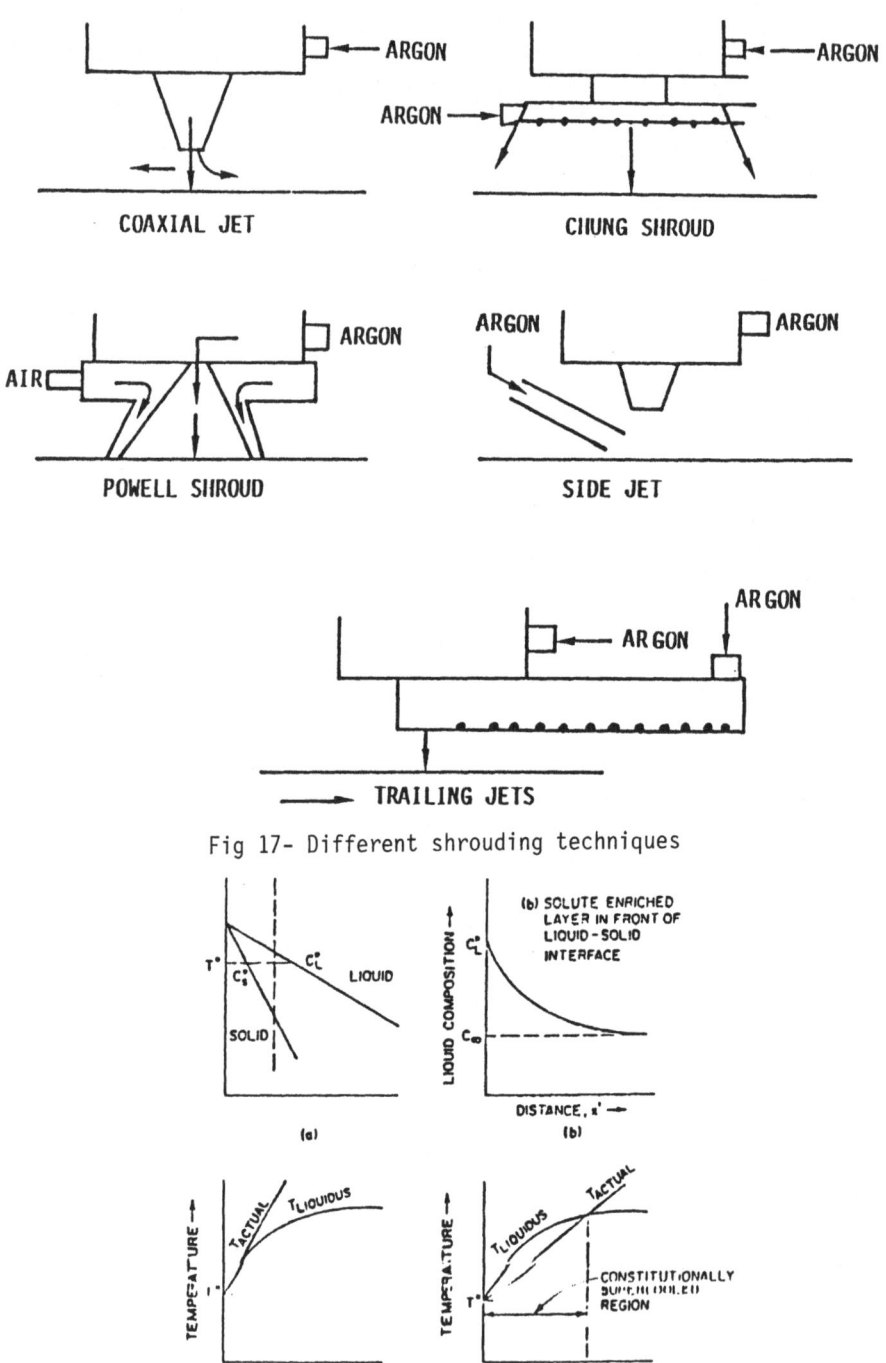

Fig 17- Different shrouding techniques

Fig. 18- Constitutional supercooling in alloy solidification. (a) Phase
diagram; (b) soluteenriched layer in front of liquid-solid inter_
face; (c) stable interface; (d) unstable interface (8)

stirred tank whose composition will be determined by:

$$D_L \frac{\partial^2 C_L}{\partial y^2} = \frac{\partial C_L}{\partial t}$$

Now:

$$\partial C_L / \partial t = (\partial C_L / \partial t)(\partial t / \partial x)(\partial x / \partial t) = - GR/M_L$$

Substituting and integrating we get:

$$\left(\frac{\partial C_L}{\partial y}\right)_{y=0} = - \frac{G R 1}{M_L D_L} \quad \text{and} \quad \Delta C_L{}_{MAX} = - \frac{G R 1^2}{2 M_L D_L}$$

From this we can see that the parameter (GR) is related inversely to the cell spacing, 1 . (GR) is the cooling rate in $^\circ$C/s. In laser surface melting extremely high cooling rates can be achieved (approx 10^6 C/s).

Heat Flow in Melting:

If the laser melting process is to be understood it is necessary to quantitatively evaluate:

* G,R for solidification reasons:
* The thermal gradient and time at temperature for stress reasons as well as diffusion or dissolution effects:
* Total thermal history of the melt for surface tension effects and time for the escape of bubbles.

This mathematical modeling can be done in several ways:
1. One dimensional heat flow: (9)

Heating up:

$$T(z,t) = \frac{\varepsilon 2 F_o}{k} (kt)^{\frac{1}{2}} \text{ ierfc}\left(\frac{z}{2(kt)^{\frac{1}{2}}}\right)$$

where: $T(z,t)$ = temperature at time, t, and depth, z.
ε = emissivity; F_o = average power density W/cm^2
k = thermal conductivity W/u$^\circ$k ; K = thermal diffusivity m^2/s

Cooling down:

$$T(z,t) = \frac{2 F_o k^{\frac{1}{2}}}{k}\left[t^{\frac{1}{2}} \text{ ierfc}\left(\frac{z}{2(kt)^{\frac{1}{2}}}\right) - (t_o-t_1)^{\frac{1}{2}} \text{ ierfc}\left(\frac{z}{2(k(t-t_o))}\right)\right]$$

2. Stationary Gaussian surface source: (10)

$$T_{\substack{cont \\ gauss}}(0,0,t) = \frac{2\,Q\,d}{\pi\,d^2\,k\sqrt{\pi}}\,\tan^{-1}\left(\frac{4\,\alpha\,t}{d^2}\right)^{\frac{1}{2}}$$

where Q = Heat input, W

This gives a useful rule of thumb for calculating the maximum surface temperature:

$$\frac{T\,\pi\,Kd}{P\,(1-r_F)} = \sqrt{\pi} = 1.77$$

3. Moving Point Source:

Rosenthal solution (11): (Swifthook and Gick (12))

$$T-T_o = \frac{Q}{2\,\pi\,k}\,C^{-\frac{vx}{2\alpha}}\,\frac{C^{-\frac{vR}{2\alpha}}}{R}$$

where v = transverse speed
α = thermal diffusivity
$R = \sqrt{x^2 + y^2 + z^2}$

Thermal Gradient: (cooling rate)

$$\frac{\partial T}{\partial t} = 2\,\pi\,k\left[\frac{v}{Q}\right](T - T_o)^2$$

Easterling and Ashby (13,14) developed an interesting modification of this by assuming the point source was sited a small distance above the surface. This then produced a fairly simple mathematical model of a moving "Gaussian Source".

5. Overall Heat Balance:

From Swifhook and Gick's (12) analysis of the Rosenthal method they derived a limiting melting efficiency for high speed surface melting of 48%.

Thus under these conditions the maximum melt zone will be given by:

$$0,48\,P = \rho\,udwS\,(C_\rho T_m + L_m)$$

where: P = Incident absorbed power W
  ρ = Density   Kg/m
  u = Traverse speed   m/s
  d = Depth of melt zone m
  w = width of melt zone m
  S = shape factor
  C_ρ = Specific Heat   J/Kg C
  T_m = Melting point   C
  L_m = Latent Heat of Fusion J/Kg

5. Finite difference models (15,16)

Although these types of solution require a large computer for the storage of the element values, they are currently the only methods available which truly explore the physics of the melting process. They are capable of modeling the heat balance on many small elements. They can thus model heat conduction in the bulk, heat convection from the surface, heat radiation from the surface melting with latent heat effects, variations in thermal properties with temperature and variations in surface reflectivity.
Some results are seen in fig. 20, 21 and 22.

Material Flow within the melt pool:

There are many forces acting on the melt pool as shown in Fig. 23 (29).

One of the largest is that from the variation in surface tension due to the steep thermal gradients:

Consider the example of Fig. 24.

$$\text{Surface Shear force} = \mu \frac{du}{dx} = \frac{\partial \sigma}{\partial x} = \frac{\partial \sigma}{\partial T} \frac{\partial T}{\partial x}$$

For Ni $\dfrac{\partial \sigma}{\partial t}$ = 0.38 ergs/cm^2 °K

For Laser Processing $\dfrac{\partial T}{\partial t}$ = 2.5x10^4 C/cm

Therefore Shear Force = 0.38x2.5x10^4 ≈ 10^4 dyns/Cm
           ≈ 10G!

Some preliminary work on modeling the flow has been undertaken by Mazumder (17). His calculations suggest that in his example the melt pool rotates around 5 times before solidifying. The marker experiments of Takeda (44) reported under cladding also indicate a very rapid mixing taking place within the melt pool.

162

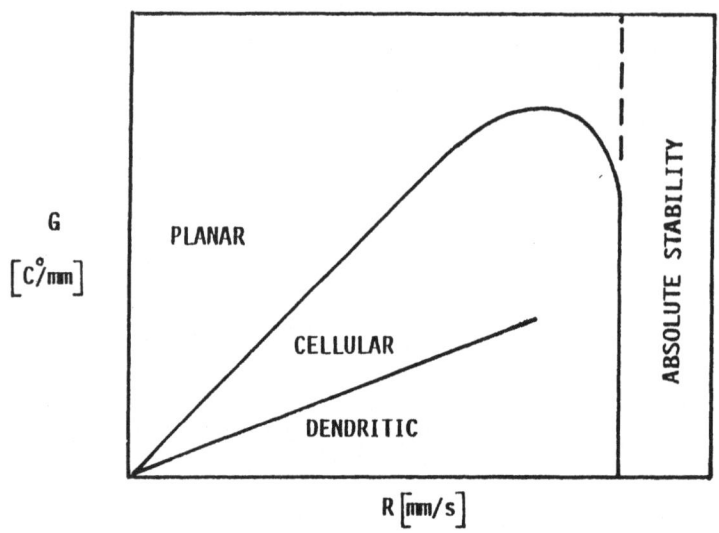

Fig·19- Morphological regions of solidification

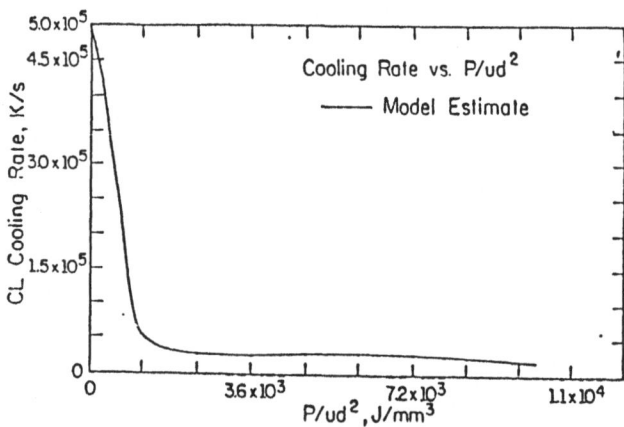

Fig 20- Model estimate of melting process

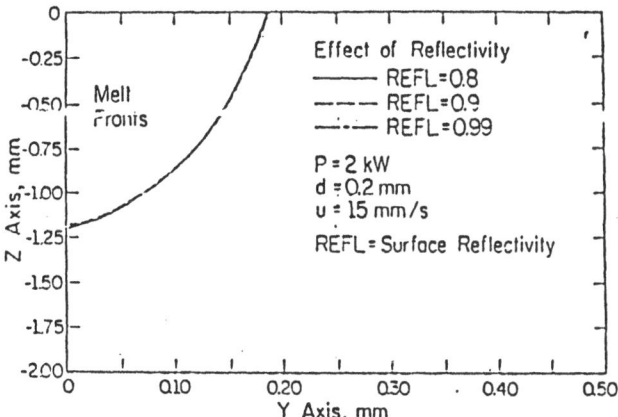

Fig 21- Model estimate with surface reflectivity effects

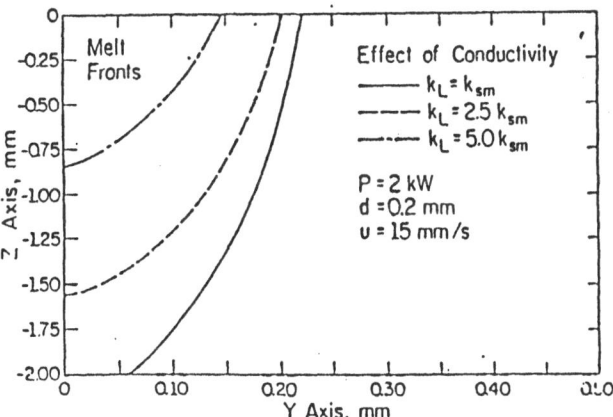

Fig 22- Model estimate with effects of conductivity

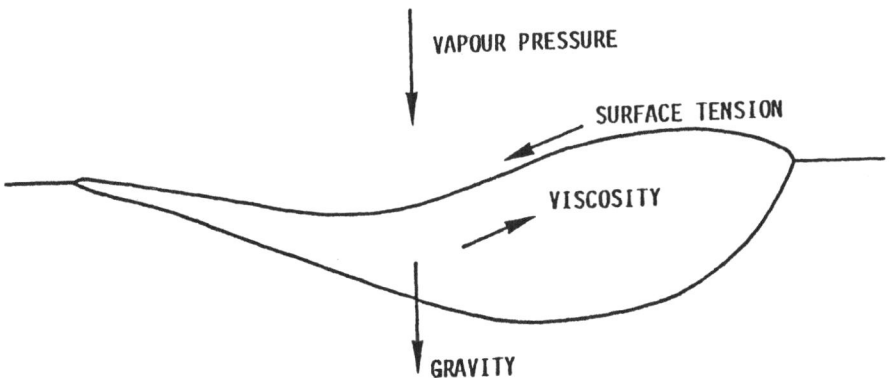

.Fig 23- Forces on a melt pool

164

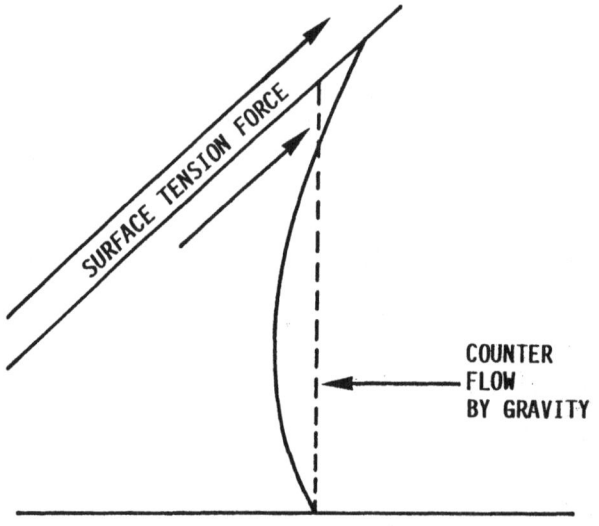

´Fig 24- Modeling of material flow within the melt pool

Product:

Cast irons: Cementite/ledeburite structure with interesting diffusion effects at the edges and HAZ. ref: 18-21

Stainless Steel: Martensitic or Austenitic fine structures. Residual tension in the austenitic steels and compression in single tracks of martensitic steel which become tensile if overlapped. The tension affects the corrosion properties. ref: 22-28.

Titanium: Highly dislocated fine structure. ref 30-34

Tool and special steels: Very hard fine carbide dispersions with high hot hardness, ref 35-38.

Bronzes: ref: 39-40

Glassy Metals: ref: 41-42

In all materials there is a tendency to cracking if the hardness is high. Usually this can be avoided if some preheat is applied.

References:

1. Olsen. F.O. "Laser Material Processing at the Technical University of Denmark" Proc Materialbearbeitung mit CO_2 - Hochleistungslasern Stuttgart April 1982 paper 3.

2. Jorgensen. M. Met Costr. Feb 1980 12 (2) 88 1980

3. La Rocca. A.V. "Laser Applications in Manufacturing" Scientific American March 1982 80-87.

4. Weerasinghe. V.M., Steen. W.M. "Laser Cladding by Powder Injection" Proc conf. "Laser in Manufacturing 1" publ. IFS publ. Ltd., Kempston Bedford ed M. Kimmitt Nov 1983

5. Man. C. Ph. D. Thesis London University 1984

6. Powell. J. Ph.D.Thesis, London University 1983

7. Bell. T., Morton. P., Bergman. H.W. "Laser Surface Melting of Titanium Alloys" Proc. "Laser 85 opto-electronik" conf. Munich July 1985 paper 111.9.7

8. Flemings. M.C. "Solidification Processing" McGraw Hill Book Co. 1974

9. Gregson. V.G. "Laser Heat Treatment" ch.4 Laser Material Processing ed. M. Bass publ North Holland Publ. Co. Ltd. Amsterdam 1983.

10. Ready. J.F. "Effects of High Powered Laser Radiation" Academic Press New York 1971

11. Rosenthal. D. Trans ASME 48 848 1946

12. Swifthook. D.T., Gick. E.E.F. Weld. J. 492S 1973

13. Ashby. M.F., Easterling. K.E. "The Transformation Hardening of Steel Surfaces by Laser Beams" Acta Metallurgica to be published 1985

14. Wen Bin Li Ph.D. Thesis Lulea University Sweden 1984

15. Mazumder. J.M., Steen. W.M. "Heat Transfer Model for CW Laser Processing" J.Appl.Phys. Feb (1980) 51 (2) pp 941-946

16. Steen. W.M., Mazumder. J. "Mathematical Modeling of the Laser/Material Interaction" GRA vol 84 N. 17 Aug 17th 1984.

17. C. Chan., Mazumder. J., Chen. M.M. "A Two Dimensional Transient Model for Convection in Laser Melted Pool" Met Trans A Vol 15A Dec 1984 pp 2175-2184.

18. Nilsson. Y. "Surface Melting of Cast Iron with a High Powered Laser Beam" Rapidly Sol.. Amor. and Cryst. alloys publ: Elsevier Science Publ: CO. New York 1982 pp 517-521.

19. Blarasin. A., Corcoruto. S., Belmondo. A., Bacci. D. "Development of a Laser Surface Melting Process for the Improvement of the Wear of Gray Cast Iron" Wear 86 (2) pp 315-325 1983

20. Trafford. D.N.H., Bell. T., Legaw. J.H.P.C., Bransden. A.S. "Laser Treatment of Grey Iron" Metals Technology Vol 10 N.2 pp 69-77 1983

21. Hawkes. I.C., Walker. A.M., Steen. W.M., West. D.R.F. "Application of Laser Surface Melting and Alloying to Alloys Based on the Fe-C System" Lasers in Metallurgy 2, Los Angeles Feb 1984 ed. K. Mukher Jure, J. Mazumder ASM publ 1984

22. Moore. P.G. "Opportunities for Surface Modification Technology in Conservation of Chromium" Tech Aspects of Crit: Mat: use in steel Vol II B publ Nat Bur Standards p 5 1983

23. Molian. P.A., Wood. W.E. "Non Equilibrium Phases in Laser Processed Fe 0.2 wt% C, 20 wt% Cr Alloys" Mat: Sci Eng: Sept 1983 60 (3) 241-245

24. Molian. P.A., Wood. W.E. "Ferrite Morphology in Rapidly Solidified Ferritic Fe-Cr-C Steels" Scr: Metall: April 1983 17 (4) 431-434

25. Lamb. M. Ph.D. Thesis London University 1985

26. Lamb. M., Man. H.C., Steen. W.M., West. D.R.F. "The Properties of Laser Surface Melted Stainless Steel and Boronised Mild Steel" Proc CISFFEL Lyon Sept. 1983 Vol 1 pp 227-234 publ by Le Commissariat a l'Energie Atomique, France

27. Lamb. M., Steen. W.M., West. D.R.F. "Structure and Residual Stresses in Two Laser Surface Melted Stainless Steels" Proc conf. "Stainless Steel '84", Gothenburg, Sweden Sept 1984

28. Vitek. J.M., Dasgupta. A., David. S.A. "Microstructural Modification of Austenitic Stainless Steels by Rapid Solidification" Metall: Trans. A Sept 1983 14 A (9) 1833-1841

29. I.C. Hawkes, Lamb. M. Steen. W.M., West. D.R.F. "Surface Topography and Fluid Flow in Laser Surface Melting" Proc CISFFEL Lyon France Sept 1983 vol 1 pp 125-132 publ by Le Commissariat a l''Energie Atomique, France

30. O'Neal. J.E., Sastry. S.M.L., Peng. T.C., Tesson. J.F. "Microstructures of Rapidly Solidified Titanium Alloys" Microstructural Science Vol 11 Orlando Flor. USA pp 18-21 July 1982

31. O'Neal. J.E., Peng. T.C., Sastry. S.M.L. "Microstructural Studies of Laser-Melted Titanium-Rare Earth and Titanium-Boron Alloys" Elect: Micrs: Soc of America 39th Annual Meeting Atlanta Geog: USA 10-14 Aug 1981

32. Konitzer. D.G., Muddle. B.C., Fraser. H.L. "A Comparison of the Microstructure of a Cast and Laser Surface Melted Ti-8A1-4Y" Metall Trans A Oct 1983 14 A (10) 1979-1988

33. Sastry. S.M.L., Peng. T.C., Meschter. P.G., O'Neal. J.E. "Rapid Solidification Processing of Titanium Alloys" J.Met 35 (9) pp 21-28 Sept 1983

34. Folkes. J., Henry. P., Lipscombe. K., Steen. W.M., West. D.R.F. "Laser Surface Melting and Alloying of Titanium Alloys" Proc 5th Int conf on Titanium, Munich Sept 1984

35. Carbucicchio. M., Meazza. G., Palombarini. G., Sambogna. G. "Surface Melting of a Medium Carbon Steel by Laser Treatment" J.Mater: Sci. May 1983 18 (5) 1543-1548

36. Lewis. B.G., Gilbert. D., Strutt. P.R. "Microstructure Control by Electron Beam Surface Melting Techniques" Proc Sym 'Processing and Properties of High Speed Tool Steels' ed. Wells.M.G.H., & Lherbier. L.W. publ AIME Warrendale Pa. 1980

37. Molian. P.A. "Structural Characterization of Laser Processed Molybdenum Steel" Mat: Sci: and Eng: Vol 58 N.2 April 1983 pp 175-180

38. Christodoulou. G., Walker. A.M., Steen. W.M., West. D.R.F. "Laser Surface Melting of some Alloy Steels" Met Tech. June 1983 10 (6) pp 215-223

39. Sekhar. J.A. "Rapid Solidification of Alloy Substrates by Laser and Electron Beam: Heat Flow Modeling and Solidification Morphology" Diss: Abs Int May 1982 42 (11) p 248

168

40. Draper. C.W., Vanderberg. J.M., Preece. C.M., Clayton. C.R. "Characterization and Properties of Laser Quenched Aluminium Bronzes" Rap: Sol: Amorph: and Cryst: Alloys pp 529-533 publ Elsevier Science Publ: Co. Inc. New York 1982

41. Hillert. M. "The Thermodynamics of the Glass Transition" Rapidly Solidified and Amorphous and Crystalline Alloys ed Kear Giessen Cohen pp 3-13 Publ North Holland publ. Co. 1982

42. Bergmann. H., Mordike. B.L. "Production and Properties of Amorphous Layers of Metal Substrates by Laser and Electron Beam" ibid pp 497-503 1982

43. Bramson. M.A. "Infra Red Radiation: A Handbook for Applications" Plenum Press, New York 1968.

44. Takeda. T., Steen. W.M. "Laser Cladding with Mixed Powder Feed" Proc conf 'ICALEO '85 ' Boston Nov 1984.

III. Laser Surface Alloying (LSA)

Summary:

        Description of the process
           experimental arrangement
           main characteristics

        Process Variations
           alloy delivery
           reflectivity
           optical
           shrouding
           applications

        Comparison with alternative processes

        Process Mechanisms
           flow in melt pool; diffusion

        Product

        References

Description of the Process:

Surface Alloying with a laser is similar to surface melting with the laser except that another material is added to the melt pool. It is also similar to surface cladding in that cladding process is performed with insufficient material for the power delivered then surface alloying would result. It is thus one extreme of surface cladding.

Experimental arrangement:

The experimental arrangement is similar to that for surface melting. The difference lies in the method of delivering the alloy material. A fairly high power density is required and gas shrouding is necessary in most situations.

Main Characteristics:

* The alloyed region shows a fine microstructure with nearly homogeneous mixing through out the melt region. Only in very fast melt tracks (~0.5m/s) can inhomogenieties be seen. Some apparent compositional differences, which show by different etch rates, are in reality only different sections through the dentrites in the fusion zone.
* Most materials can be alloyed into different substrates. The high quench rate ensures that segregation is minimal. In fact some surface alloys can only be prepared via a rapid quench; e.g. Fe/Cr/C/Mn (1).
* Thickness of treated layer can be from 1-2000 μm. Very thin, very fast quenched alloy regions can be made using Q switched lasers.
* Some loss of the more volatile components can be expected (2)

*   Other characteristics are as for surface melting

Process variations:

Alloy delivery systems:

    The alloy can be place in the melt zone by:
       1. Electroplating (3)
       2. Vacuum Evaporation
       3. Preplaced powder coating (4)
       4. Thin foil application
       5. Ion Implantation
       6. Diffusion e.g. Boronising (5,6)
       7. Powder blowing (7)
       8. Wire feed
       9. Reactive gas shroud; e.g. $C_2 H_2$ in Ar or just N_2 (8,9)

Reflectivity: Optical Arrangement: Shrouding:

All these processing conditions are similar to that for surface melting which was described previously.

Applications:

    The potential applications are numerous for a process which could produce almost any surface alloy required, can be engineered as a non contact method and can be automated.
    The alternatives open to design engineers by this process are so many that they represent a real challenge to our metallurgical knowledge. The main question facing design engineers is "What alloy would I like?".
    In 1982 Draper wrote in his review (10) "Laser surface alloying is far from maturity. In fact it could hardly be placed in its adolescence". Today three years later one could write similarly. There has been an explosion of scientific papers on the subject but they only represent a mountain of information when there is still the Pyrenes to explore. Nevertheless this activity has been seriously directed towards industrial applications and many processes are now being considered with detailed care.

Comparison with alternative Processes:

The main competing processes are:

Cladding and plating:
    These processes offer better control over depth, and composition. They also offer the advantages of totally different alloys for corrosion or wear protection. However the cladding which can be laid down by fusion bonding is limited to alloys which are compatible for welding; similarly not all materials can be plated successfully.

Boronising; Carburising; Chromising:
    These diffusion processes are performed in the solid state and

therefore have no surface flow. However the long processing times at high temperatures which are required often leads to considerable distortion even though the surface finish may not be unduly affected.

Laser Surface alloying offers precision of the placement of the alloy and vastly improved processing speeds.

Process Mechanisms:

Flow in Melt Pool:

The very good mixing which is observed in LSA tracks indicates that intense stirring of the pool is occurring. The diffusion rate was calculated by Mazumder and Chande (11) to be around 10^8 cm^2/s larger than the expected molecular diffusivity within the liquid state. Thus strong eddy diffusion must be occurring. The mechanism for this has been described previously.

Defects: (14)

These are mainly stress or liquation cracking. There is also a risk of porosity particularly with rimming steels as substrates.

Product:

Titanium:

The reactivity of Titanium makes it an ideal target for laser surface alloying by gas reaction. Studies by Bergmann (8) and Folkes (9) show that if C is alloyed into Ti then TiC forms as the primary dendrites giving a very hard matrix which will not drop out as injected particles might. In alloying in a Nitrogen atmosphere TiN is formed and due to the variation of melting point with concentration simultaneous solidification of primary TiN dendrites from the top and bottom of the melt are observed. Preliminary work by Bergmann shows improved fatigue properties for TiN and many workers have observed the wide range of surface colors which can be obtained this way. Titanium art work is becoming well know and much respected. It may be that the laser has something to offer in this area.

Cast Iron: Surface alloying with Cr, Si, and C are all possible and are currently under investigation by Zhen Da (12).

Steel: a study by Christodoulou (3) on Chromium electroplated special alloy steels showed the wide range of alloy compositions obtainable; but also the great need for a very stable power delivery system. Interesting in this study Christodoulou was able to generate coatings in which the top surface of the plating had melted and the interface region, but there was a band of unmelted Cr in between. This is a possible method of sealing Chromium plate.

Other systems explored are:

    Mo in steel (16)
    B   "   "   (5,6)
    Cr "   "   (3,13,17,18)
    Ni "   "   (15,18)

Stainless Steel: C in Stainless Steel (19)

Aluminium: This has been successfully alloyed with Si, C, N by Walker et al

(21) and Picraux (22).

Superalloys: Cr in Superalloys (20)

References:

1. Mazumder. S. "Wear Properties of Laser Alloyed Fe-Cr-Mn-C Alloys" ICALBO '84 Vol 34 Boston 1984

2. Blake. A., Mazumder. J. "Control of Mg Loss During Laser Welding of Al-5083 Using a Plasma Suppression Technique" ASME J. of Eng. for Ind. to be published Aug 1985

3. Christodoulou. G., Steen. W.M. "Laser Surface Treatment of Chromium Electroplate on Medium Carbon Steel" Proc 4th Int conf on Laser Processing, Los Angeles Jan 1983

4. Walker. A.M., West. D.R.F., Steen. W.M. "Laser Surface Alloying of Ferrous Metals with Carbon" Proc Laser '83 Optoelectronik conf, ed. W. Waidelich, Munich June (1983)

5. Man. C.Ph.D. Thesis London Univ. 1984

6. Lamb. M., Man. C., Steen. W.M., West. D.R.F. "The Properties of Laser Surface Melted Stainless Steel and Boronised Mild Steel" Proc CISFFEL Lyon Sept 1983 pp 227-234 publ by Le Commissariat a l'Energie Atomique, France.

7. Weerasinghe. V.M., Steen. W.M. "Laser Cladding with Pneumatic Powder Delivery", Proc 4th Int conf. on Laser processing, Los Angeles Jan (1983)

8. Bergmann. H.W., Bell. T., Lee. S. "Thermochemical Treatment of Titanium Alloys with Lasers (Laser Gas Alloying)" Laser '85 Munich 1985

9. Walker. A.M., Folkes. J., Steen. W.M., West. D.R.F. "The Laser Surface Alloying of Titanium Substrates with Carbon and Nitrogen" Surface Engineering Vol 1 Jan 1985

10. Draper. C.W. "Laser Surface Alloying: The State of the Art" Jour Metals June 1982 pp 24-32

11. Chande. T., Mazumder, "Mass Transport in Laser Surface Alloying: Ion-Nickel System" J. Appl.Phys Letters Vol 41 N.1 p 42 1982

12. Zhen Da.C. private communication

13. Molian. P.A., Khan. K.H., Wood. W.E. "Microstructure of Laser Processed Fe-Cr Surface Alloys" Electron Microsc: Soc Am: 39th Meeting Atlanta Geo USA 10-14 Aug 1981

14. Molian. P.A., "Characterization of Fusion Defects in Laser Surface Alloying Applications" Scripta Metallurgica Vol 17 N. 11, pp 1311-1314 1983

15. Chande. T., Mazumder. J. "Compositional Control in Laser Surface Alloying" Met Trans B Vol 14B N.2 June 1983 pp 181-190

16. Tucker. T.R., Clauer. A.H., Ream. S.L., Walkers. C.T. "Rapidly Solidified Microstructures in Surface Layers of Laser Alloyed Molybdenum on Fe-C Substrates" Proc Conf Rapidly Solid: Am: + Cryst: Alloys Boston Mass Nov 1981 pp 541-545 publ Elsevier Science Publ Co Inc. New York 1982.

17. Mordike. B.L., Bergmann. H.W. "Surface Alloying of Iron Alloys by Laser Beam Melting" ibid pp 463-483

18. Lumsden. J.B., Gnanamuthu. D.S., Moores. R.J. "Corrosion Resistance of AISI 4140 Steel Laser Surfaces Alloyed with Cr and Ni" Proc Conf Corrosion of Metals Proc by Energy Beams Louisville Ken USA 13th Oct 1981 pp 129-134

19. Marsden. C., Steen. W.M., West. D.R.F. "Laser Surface Alloying of Stainless Steel" to be publ. NATO Inst San Miniato 1985

20. Tien. J.K., Sanchez. J.M., Jarrett. R.T. "Outlook for Conservation of Chromium in Superalloys" Proc Tech: Aspects of Critical Materials use by the Steel Industry Vol 11-B Nashville Tenn USA 4-7 Oct 1982 p 30 Publ. Nat Bur Stds, Washington USA 1983

21. Walker. A.M. Ph.D. Thesis London University to be published.

22. Picraux. S.T., Follsteadt. D.M. "Surface Modification and Alloying: Aluminium" Report N. DE82018440 Sandia Nat: Lab: p 70 May 1982

IV - Laser Surface Cladding (LSC)

Summary:

Description of the process:

Process Variations:
    Mixed powder feed
    reflectivity
    vibro cladding
    applications

Comparison with alternative processes:

Process mechanics:
    dilution
    pool shape
    stress, cracking, porosity

Product:
    microstructures   homogeneity
    properties   wear
                      corrosion
                      adhesion
                      mechanical strength

Description of the Process:

The objective in laser cladding is to fuse an alloy on to the surface of a substrate with the minimum of dilution with the substrate. Areas are clad by overlapping single clad tracks.

The process can be done by either preplacing a powder on the substrate, or blowing the powder into the laser generated melt pool. Or it can be done applying the clad material as a wire, sheet or plasma spray coat or electroplate coat.

In preplace powder cladding described by Powell (1) the laser sends a melt wave through the powder bed. Since the powder bed has a low thermal conductivity the pool is almost thermally insulated until it reaches the substrate surface. At that moment it will freeze back forming only a solid/liquid bond which is relatively weak compared to a full fusion bond. Continued heating will remelt the back solidified material and then cause a fusion bond to form. The process is illustrated in Fig. 25. One of the principle problems with this process is the difficulty in keeping the powder in place while it is melted by the beam.

In blown powder cladding (2,3) the powder is blown in an inert gas stream into the laser generated melt pool as illustrated in Fig. 26. The leading edge of the melt pool will incorporate the substrate. Particles arriving in this area will be solid. If the leading edge of the substrate is also solid then the particle will not stick and cladding will not occur. If however the leading edge is molten then the particle will stick and will melt almost instantly under the power from the beam, thus forming a fusion bond. The level of dilution is controlled by the powder feed rate

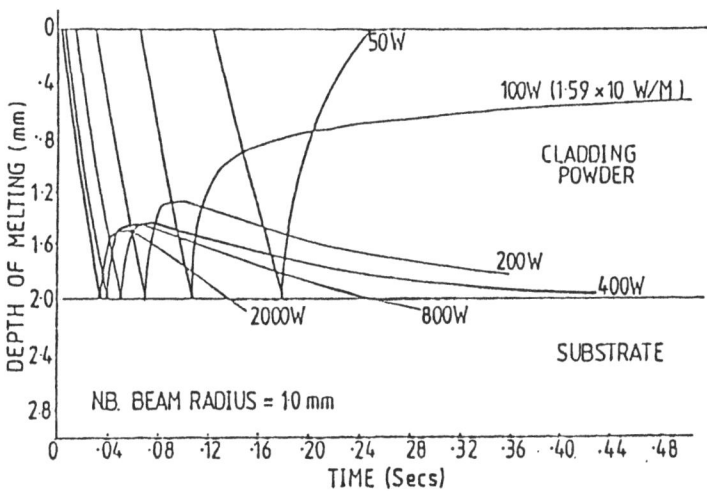

Fig 25- Movement of molten front with time at different
laser powers (1)

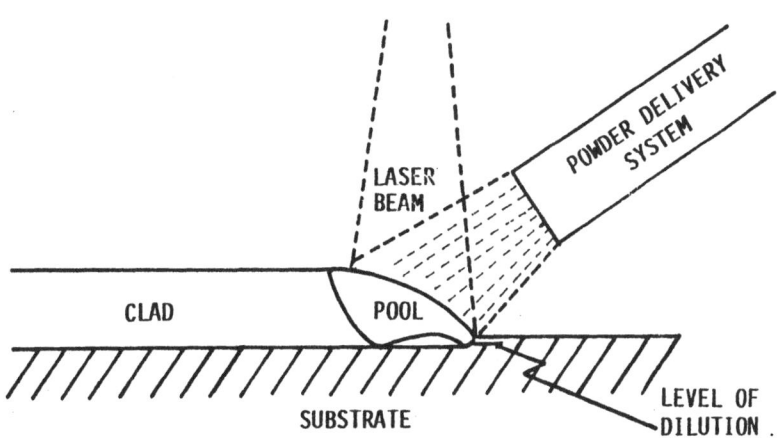

Fig 26- Laser cladding with blown powder

determining the size of the substrate's molten leading edge.

If the clad material is fed as either wire or sheet there is a problem with reflectivity and so the process is less efficient but still possible.

Consolidation of surfaces by laser remelting of plasma spray coated or electroplated surfaces is also practiced. (4,5)

The main Characteristics of the blown powder process are:

* Controlled levels of dilution
* Localized heating which reduces thermal distortion
* Controlled shape of clad within certain limits.
* Smooth surface finish (25 μm)
* Good Fusion bonding
* Fine quench microstructures.
* Non Contact method of application
* Easily automated
* Omnidirectional.

Process variations:

* Mixed powder feed: (6) By this method alloys can be formed in-situ or non homogeneous deposits formed.

* Optical feed back systems (2,3) have been developed which increase the efficiency of utilization of the beam power by around 40%.

* Vibro Laser cladding: (7) By this method the substrate is vibrated ultrasonically while cladding proceeds. There is considerably less cracking and porosity observed.

Applications:

The first industrial use of lasers in cladding was done by Rolls Royce (8) in 1981. They clad turbine blade shroud interlocks on the RB211 engine. Since then many companies are applying or considering applying this process.
Eboo (9) lists the following activity:

Production Stage

Company	Component	Comment
Rolls Royce	turbine blade shroud interlock	triballoy/nimonic powder feed
Pratt & Whitney	turbine blade	PWA 694/nimonic preplaced chips
G.E.	proprietory	reverse machining with Ti powder feed

Pilot Demonstration Stage

Company	Component	Comments
Combustion Eng	offshore drilling & production parts	stellites,Colmonoys and other alloys including carbines
	valve components boiler firewall	powder feed
FIAT	valve stem valve seat Aluminium Block	Cr C,Cr,Ni,Mo/cast Fe preplaced powder
GM	automotive	cast iron systems
Rockwell	aerospace	T-800,Stellites,powder feed
Westinghouse	turbine blades	Stellites,Colmonoys preplaced beds & gravity
NRL	proprietory	multiple alloy, powder feed

Comparison with alternative processes:

One reason for the great industrial interest in this process is that it is one of the very few surfacing processes which cause little thermal damage to the substrate, and relatively slight distortion.

Alternative processes are compared in the table below:

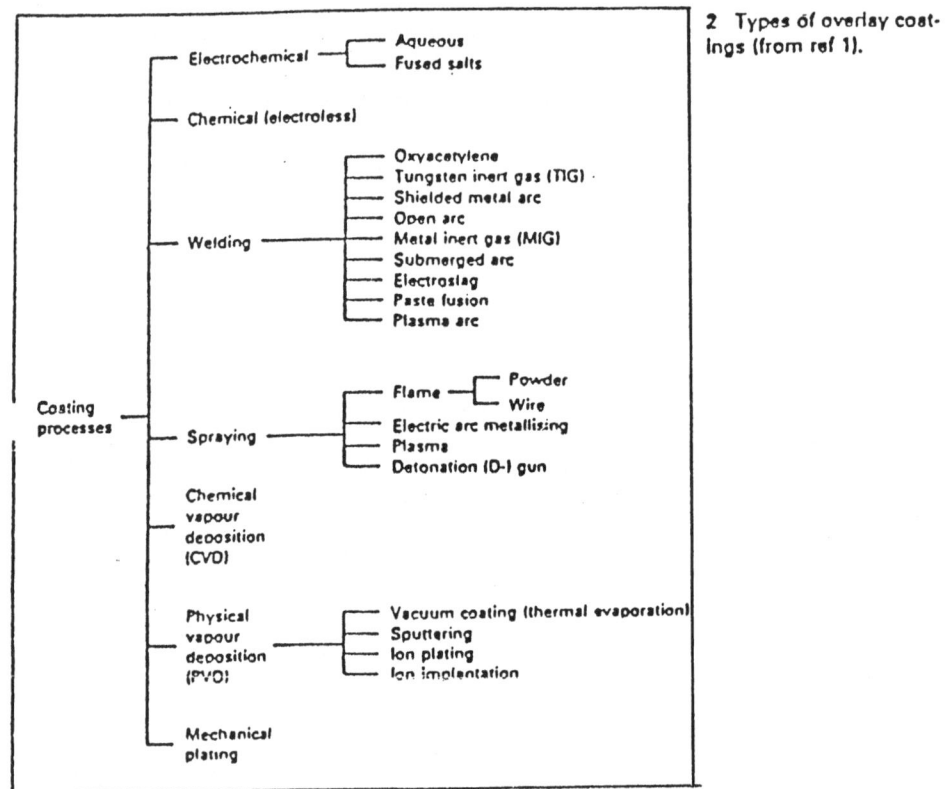

2 Types of overlay coatings (from ref 1).

Process Mechanics:

1. Dilution:
    Dilution is principally dependent upon the powder feed rate for a given specific energy. There is a fairly generous operating zone in which a good fusion bond can be achieved with only a 10-20 μm dilution zone.

2. Pool Shape:
    The pool shape and traverse speed define the molten time. This term affects homogenization in mixed powder feed systems and thermal penetration in all systems. Marker experiments undertaken by Takeda (6) and illustrated in fig 27 give the values shown in the table for the molten time and pool shape.

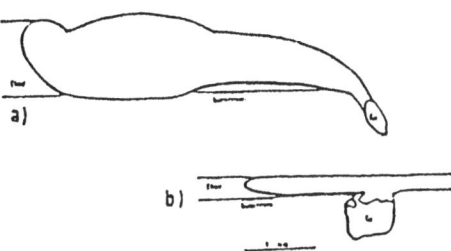

Fig 27- "Diagram showing the extent of the flow of copper into a clad
melt pool from a copper marker wire buried in the substrate.
The copper was found to be fairly uniformly distributed within
the marked zones.
The two tracks were made at different speeds:
a) 4mm/s   b) 12mm/s

Table 2. Pool sizes and molten times for various traverse speeds for a laser power of 1.86 KW 4mm beam dia and 0.315 g/s powder feed

Speed mm/s	Pool Length mm	Molten Time s	Theoretical Molten Time s
4.4	5	1.13	
7	3.5	0.5	0.3 (cf Fig. 25)
12	1.8	0.15	
19.5	1.7	0.19	

3. Stress, Cracking and Porosity:

The clad layer will normally have a large residual tensile stress within it. This is due to the restrained contraction of the prior molten clad. Only when cladding prestressed or bent substrates can a compressive stress be created in the clad track. When overlapping tracks there is a tensile 'Bow wave' ahead of the new clad this tension added to the existing residual stress may cause cracking in hard materials. Cracking can usually be overcome with preheating the substrate.

Porosity is of two types: interrun porosity in which a bubble is trapped between successive overlapping passes and porosity within the clad or at the clad interface. The interrun porosity is due to a poor shape on the individual clad track. The tracks must have an obtuse angle at their edges. Porosity at the interface or within the clad is usually due to some chemical reaction occurring, as with rimming steels and certain cast irons. An interlayer or chemical addition may be required to stop the reaction from occurring.

Product:
The product and operating conditions are described in the attached paper. (10)

References:

1. Powell. J. Proc conf on "Surface Engineering with Lasers" London May 1985 Paper 17 publ. Metal Society, London

2. Weerasinghe. V.M., Steen. W.M. "Laser Cladding with Pneumatic Powder Feed" Proc 4th Int Conf on Laser Processing Los Angeles Jan 1983

3. Weerasinghe Ph.D. Thesis London Univ. 1985

4. Dallaire. S., Cielo. P. "Pulsed Laser Treatment of Plasma Sprayed Coatings" Met Trans B Vol 13B N.3 Sept 1982 pp 479-483

5. Bhat. H., Zatorski. R.A., Herman.H., Coyle.R.J. "Laser Treatment of Plasma Sprayed Coatings" Proc. 10th Int. Conf. on Thermal Spraying Essen W.Germany pp 2-6 May 1983
publ. Deutscher Verlag fur Schweisstechnik GmbH Dusseldorf, W. Germany. 1983

6. Takeda. T, Steen.W.M., West D.R.F. "Insitu Clad Alloy Formation by Laser Cladding" Proc. LIM 2 Birmingham UK March 1985 publ. by IFS publications Ltd, Bedford UK.

7. Powell. J., Steen. W.M. "Vibrolaser Cladding" Laser in Metallurgy ed K. Mukherjee, J. Mazumder publ Met soc of AIME Warrendale PA USA pp 93-104

8. MacIntyre. M. "Laser Hardfacing of RB 211 Turbine Blade Shroud Interlocks" proc 2nd Int Conf on Applications of Lasers in Material Processing Jan 1983 Los Angeles

9. Eboo. M., Lindemanis. A.E. "Advances in Laser Cladding Technology" LIA conf, Los Angeles March 1985

10. Weerasinghe. V.M., Steen. W.M. "Laser Cladding" to be publ Materials Tech 1985

LASER CLADDING WITH PNEUMATIC POWDER DELIVERY

V.M. Weerasinghe and W.M. Steen

Department of Metallurgy, Imperial College, London

ABSTRACT

A process of laser cladding using Argon blown powder delivery is described. The effect of the process parameters on the quality of the clad layer is discussed. The flexibility and versatility of this process are noted together with the significant process improvement obtained using a reflective shroud and shot blasted specimen.

INTRODUCTION

The use of lasers for surface modification has lagged far behind its use for cutting, welding and related work. This is largely due to practical limitations imposed by the laser. Namely, one may induce extreme temperatures via high intensity heating only at the expense of accepting small working volumes.

Cladding and surface alloying are two of the many material processing applications for which a laser has been and is being used, (ref 1-7). Use of a laser beam as a heat source is most attractive due to:

(a) 'clean' optical energy
(b) localised high intensity heating
(c) flexible handling capability
(d) stable, calm interaction with workpiece

Soares, O.D.D., Perez-Amor, M. (eds), Applied Laser Tooling. ISBN-13: 978-94-010-8096-5
© *1987. Martinus Nijhoff Publishers, Dordrecht.*

The reader, if not familiar with lasers, may visualise a beam of electro-magnetic radiation similar in many ways to ordinary light in that it can be reflected, focussed using mirrors, lenses and which can be used to heat, melt or vapourise any material on earth within limitations imposed by the specific laser.

Some advantages of laser cladding which had come to light during the course of the present study are listed below:

(a) Process flexibility
(b) Fast through put time, less labour intensive, ease of automation
(c) The major part of the cost is the capital charge and not the cost of processing
(d) Non-contact method of application
(e) Minimum surface preparation
(f) Good fusion bond
(g) Low dilution levels (0)
(h) Fine quench microstructure - good mechanical properties, less segregation effects, short homogenisation heat treatment
(i) Good surface finish
(j) Minimal diffusion of substrate elements due to short heat cycle
(k) Ultra-thin clad layers (0.3mm)
(l) Randomly orientated grain structure - near isotropic mechanical properties
(m) Localised heating - reduced thermal distortion, treatment of confined areas

EXPERIMENTAL PROCEDURES

The basic technique of laser cladding employed in the present study, is illustrated in fig. (1). The cladding material in the form of a fine powder is borne in a suitable inert gas and injected into the laser generated molten pool on the moving substrate surface. This technique is similar to that used by Rolls Royce in production (ref 8).

Fig (1) Laser Cladding by Power Injection.

Uniform clad layers, similar to that illustrated in figs (2a,b,c) are produced by overlapping single clad tracks.

Fig (2a)   Development of a uniform clad
layer by overlapping.

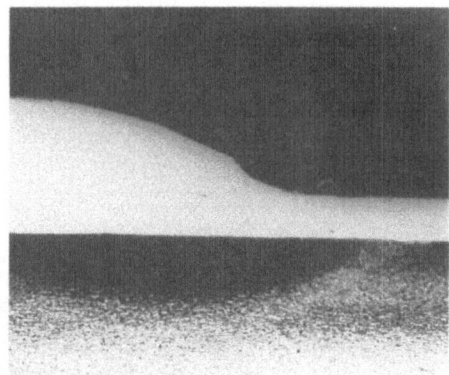

Fig.(2c) Section of Fig (2a).
First layer thickness 0.33mm.
Composite thickness 1.21mm

Fig (2b) A laser clad plate.
Base plate thickness 12.5mm
Two layers deposited

Fhe laser used was a fast axial flow type CO_2 continuous wave
machine, manufactured by Control Laser (UK) Ltd. A powder
feeder was specially designed to give a low uniform mass flow
rate (ref 9, 10). An in process powder flow monitoring system
(ref. 9) was also developed, fig (3).

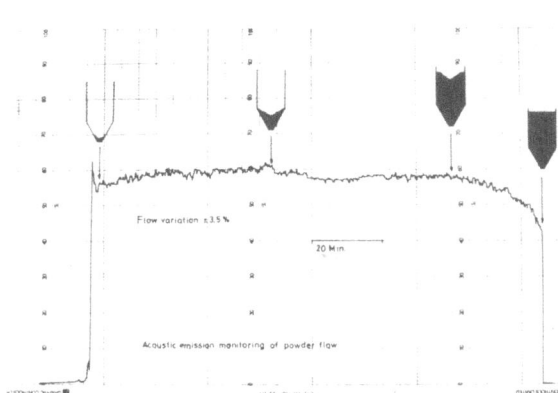

Fig. (3) In-process monitoring of powder flow using an acoustic emission technique.

The major part of the present study is related to cladding austenitic stainless 316L (16.5/18.5Cr, 11.0/14.0Ni, 2.25/3.0Mo, 0.03Max C) to a mild steel (0.2%C) flat plate substrate. Henceforth 'clad' will refer to the above stainless steel and 'substrate' will refer to the mild steel.

CHARACTERISTIC PROCESS PHENOMENA.

The powder injection technique is fundamentally more flexible and superior to the pre-placed powder technique (ref. 11) in that localised areas of components of complex geometry can be clad with better control on dilution and clad thickness.

The use of the cladding material in the form of a powder is most attractive due to the increased efficiency obtained in coupling the laser energy. Powder particles in or near the surface appear to enhance the absorption of laser energy. A molten pool is generated on a ground surface when powder is injected whereas no melting is observed without powder. Metallurgically the use of powder may reduce macro-segregation effects during solidification.

Superficial melting of the substrate surface is required so that a fusion bond takes place. The particles need not necessarily arrive at the melt pool surface in a molten state. In fact, theoretical calculations and high speed photography show that an average size particle hardly gets red hot while travelling through the beam (ref 12).

Solid particles are often observed in plasma sprayed coatings. In powder injection laser cladding, no embedded solid particles were observed in the cast structure. It has been shown theoretically that the melting time of a particle is much lower than its time of residence within the superheated melt pool (ref 9). Also, theoretical calculations show that particles may penetrate the melt pool surface only to a shallow depth before melting completely or, may ricochet from the melt surface if the impact angle became smaller than a critical angle for ricochet.

Particles falling on the 'mushy' region leave a 'crater' and as a result the surface of a clad bead appears to be 'pock-marked' fig (4).

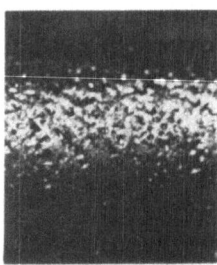

Fig (4) Characteristic 'pock-marked' surface finish of a clad bead

Normally, as much as 80% of the powder falling on the molten pool is utilised to create the clad bead. However, the overall powder utilisation will be much lower if the particles falling on the area surrounding the melt pool are also accounted. Recycling of powder is possible if there is efficient shrouding.

The process parameters are listed in Table 1 categorised into three systems; laser, powder injection and substrate handling.

Table 1 **Main operating variables for laser cladding by powder injection . (For a given clad/substrate combination).**

System Variable   Method of Monitoring

Laser - CW CO_2
      Power (up to 3kW used)   Flowing cone calorimeter
      Beam Diameter (1-5mm used) Rotating rod, (ref. 13,14)
      Mode Structure   prints, optical geometry,
                               fluorescent screens.

Powder Injection - Screw Feeder   Load cell on feed hopper
Mass Flow Rate (0.04-0.35g/s)   acoustic emission sensor
inside diameter of injector tube   on feedline
3mm)
Particle Velocity (1.4m/s for   Rotameters, fast action
conveying gas velocity of 5.8m/s)   cinephotography,
                                  photographic
                                  tracer technique

Particle Shape and Size(-100 mesh
used. Ave. size 60μm of non-   Microscopy, sieve
                                  analysis
-spherical shape)

Substrate Movement
Transverse Speed (hydraulically   Electronic timer
controlled, 5-50mm/s used)
Transverse Index (manual 0.3-2.0mm   Scale calibrated to
                                  0.1mm
used)

190

EVALUATION OF CLADDING RATES

Fig (4a,b) show the relationship between the cladding speed
and the clad bead geometry.

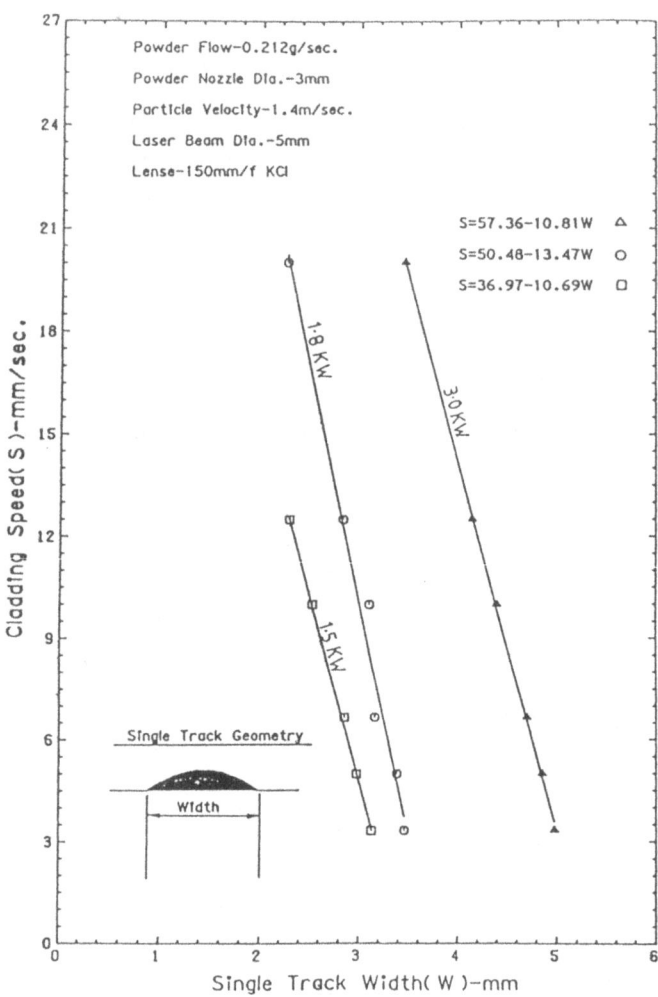

Fig (4a) Effect of cladding
speed on bead width

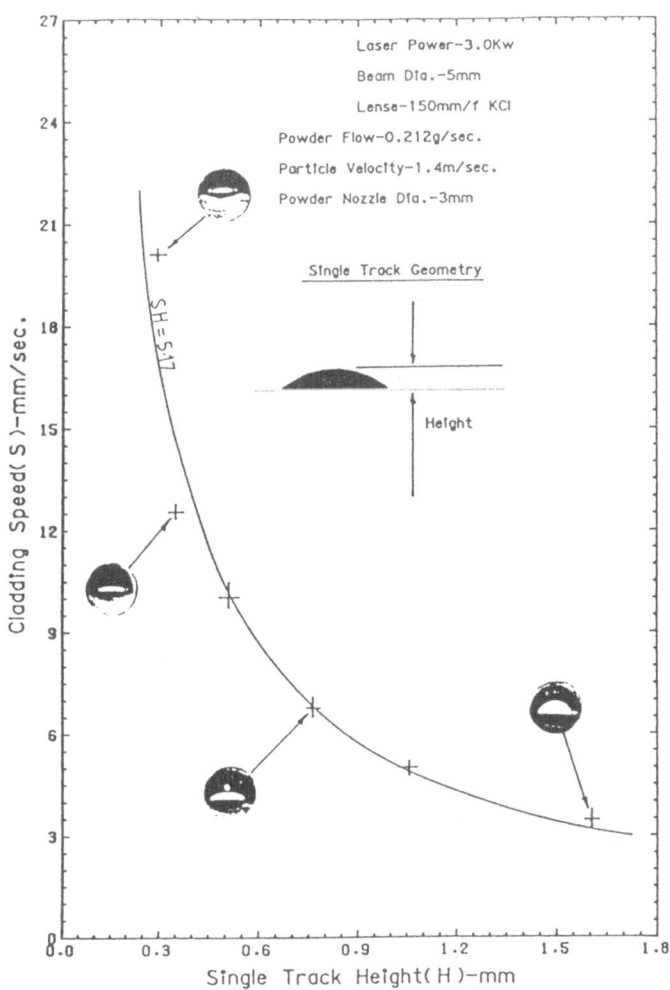

Fig (4b) Effect of cladding speed
on bead height

The effect of overlap on the clad thickness is shown in Fig (5). The percentage overlap 'k' is plotted against the dimensionless group (T/H) where 'T' is the uniform clad thickness and 'H' is the height of a single bead.

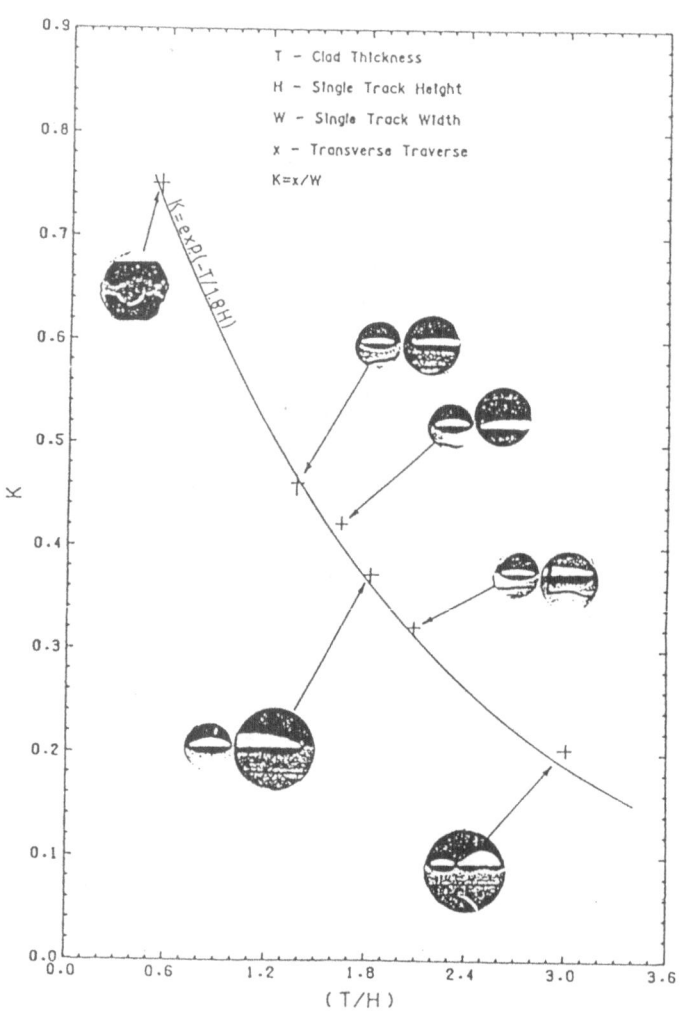

Fig (5) Effect of overlap on clad thickness

From these graphs it is possible to draw up relationships of the form:

$$S = a - bW$$

$$K = \exp(-T/_{1.8H})$$

$$SH = d$$

$$C = KWS$$

Where,

S = cladding speed
W = bead width
H = bead height
T = clad thickness
K = overlap factor as defined in Fig (5)
C = cladding rate i.e. area per unit time, assuming cladding in both directions and neglecting the time taken for transverse indexing
a, b, d = constants

It is seen that there is an optimum overlap factor 'K' for a given clad thickness 'T' since the same clad thickness can be obtained with a low 'K' and a low 'S'. The optimum 'K' is that which gives the maximum 'C' and is found by differentiation of $C = f(K, T)$. Such calculations are presented in fig (6), showing the maximum cladding rate for a particular clad thickness.

It is noted that the graphs in Fig (6) go through the origin (0,0). However, it must be stated that there is a definite region within which satisfactory cladding is obtained. Although the graphs can be extrapolated to high power levels, they cannot be safely extended to powder levels less than

194

about 1.0kW i.e. using a 5mm spot. There is a low and a high
speed limit both depending on the laser power or rather the
power density (i.e. power/spot area) and the powder mass flow.
The high speed limit is where cladding ceases due to
insufficient heat input to melt the substrate surface. The low
speed limit is where the high specific energy input distorts
the substrate plate to unacceptable levels and/or the process
becomes uneconomical.

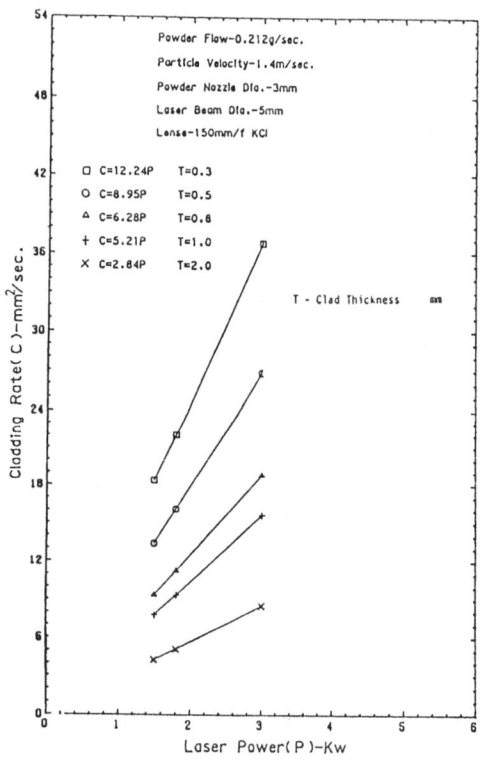

Fig.(6) Cladding coverage rates.

The cladding rates shown in Fig. (6) can be greatly enhanced
by using a hemi-spherical reflector to recycle the reflected
energy losses, Fig (7,8) and shot blasting the substrate
surface, Fig (9).

Fig.(7): Spherical reflecting
dome

Fig (8).
Effect of recycling the reflected
energy or cladding rate.

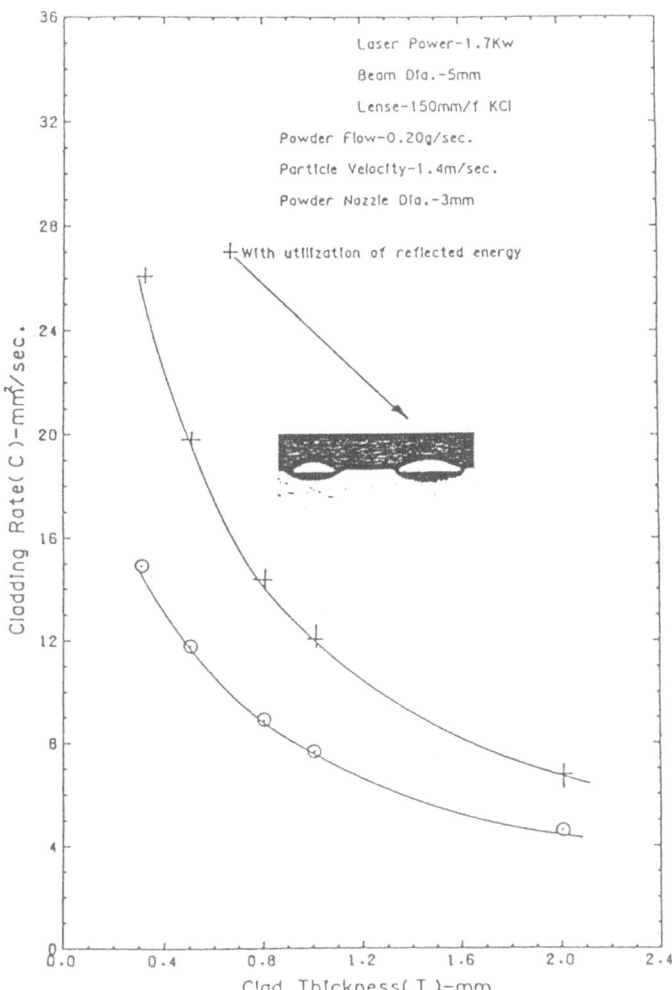

Laser Power–1.7Kw

Beam Dia.–5mm

Lense–150mm/f KCl

Powder Flow–0.20g/sec.

Particle Velocity–1.4m/sec.

Powder Nozzle Dia.–3mm

+ With utilization of reflected energy

Cladding Rate(C)–mm^2/sec.

Clad Thickness(T)–mm

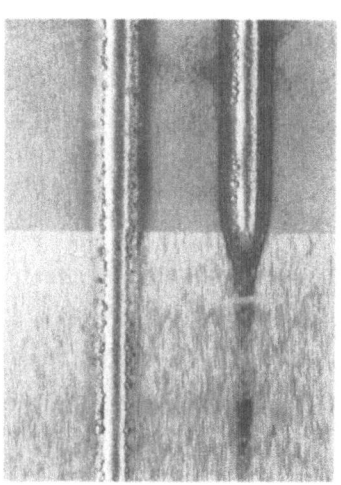

Fig (9) Effect of shot blasting the substrate surface. The top half of the substrate surface is shot blasted. The botton was masked to retain the original ground finish. The track on the left was produced with the 'dome' and continues on both sections, since the reflected energy is recycled.

By off setting the axis of the beam to the axis of the dome, the reflected beam could be positioned away from the main beam and so be used for either pre or post heat Fig (10). It also allowes an experimental determination of the reflected energy by collecting the reflected radiation in a calorimeter. Thus this device, when clean, recycles some 40% of the incident beam energy.

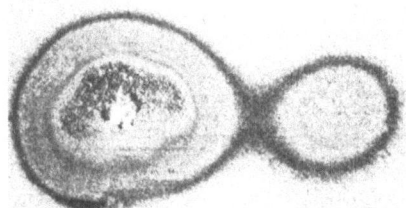 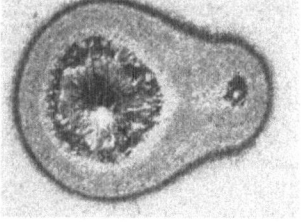

Fig (10) Reflected beam (right), positioned away from the main beam

Reference to Fig (8), it can be seen that the advantage of using the reflecting dome is greater for thinner clad thicknesses i.e. at higher cladding speeds. This is because at high cladding speeds, a small area is heated to high temperatures compared to the total area on which the laser beam is incident. Since reflectivity reduces at high temperatures, then at high cladding speeds with reduced temperatures, a higher proportion of the incident laser power will be reflected.

THREE BASIC BEAD PROFILES

Three basic single track bead profiles emerged from the numerous section profiles which were examined, Fig (11).

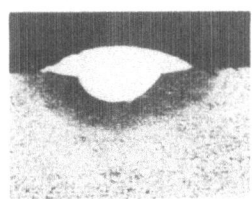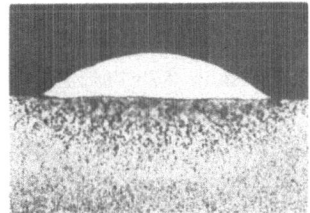

Fig (11) Three basic single track bead profiles

Profile 'c' with an obtuse contact angle is the preferred section for low dilution cladding. At higher powder feed rates or slower speeds profile 'a' is produced while at lower powder feed rates and/or higher power densities profile 'b' is produced.

Multiple track clad layers produced by overlapping of profile 'c' will be of good surface finish, minimum or no porosity and of no dilution. Profile 'a' will produce a multiple track clad layers of poor surface finish, greater thickness, no dilution and heavy porosity, especially at the root in between runs. Clad layers produced by profile 'b' will be of good surface finish, no porosity, high dilution and generally of lesser thickness than 'c'. Further the penetration pattern shown in Fig (11b) is characteristic of a Gaussian beam where there is a peak intensity at the centre. With a 'dough-nut' beam, the penetration is more pronounced at the bead edges and is generally of less severity.

EFFECT OF POWDER INJECTION PARAMETERS

The extent of dilution is controlled by the injected powder mass flow or rather the powder flux (g/smm^2). The optimum powder flux is that which gives the minimum dilution and maximum deposition and is dependent on the power density (W/mm^2) and mode structure of the laser beam and is independent of the cladding speed. For a lower powder flux the dilution will increase and mass deposition decrease. For a higher powder flux , the effect is reversed until at a certain high flux the clad track becomes continuous, Fig. 12.

The optimum powder mass flow was found to be directly proportional to (P/D x n) where 'P' is the laser power. 'D' the spot diameter and 'n' is a beam shape factor depending on the power intensity distribution.

The optimum stainless powder (-100 mesh size) flux was found to be 10mg/Smm^2 for a laser power of 1800W, spot diameter of 5mm and a 'dough-nut' intensity distribution. A 10% higher flux was required for a Gaussian beam due to its more centralised power distribution.

A 10% increase in the cladding rate was obtained when the average particle size was reduced from 77 microns to 58 microns (ref.9).

A low particle velocity ($\approx 1.4mm/s$) is desirable to minimise ejection losses from the molten pool. Since there is a minimum gas velocity required for efficient coveying ($\approx 2.8m/s$) a limit is imposed. However, some amount of gas can be purged through fine holes near the injector tube exit and so reduce the particle velocity (ref 9).

The major effect of the powder injection angle is due to the variation of the powder flux (g/smm^2) input to the molten pool. The distance from the molten pool to the injector tube has a similar effect due to divergence of the powder clad.

The powder jet is positioned in relation to the beam so that the whole of the molten pool is flooded with powder. Normally, only about 10% of the laser power is absorbed by the powder cloud. This was verified by theoretical calculation and by experiment (ref 9).

In practice, an injection angle of 45°-38° (defined from the horizontal) was used with the distance to the molten pool around 10-12mm from the injector tube end.

EFFECT OF LASER BEAM PARAMETERS

Power, spot diameter and mode structure are the beam parameters. Effects due to beam polarisation are well known in cutting (ref 15), namely preferential absorption at a certain orientation of the cut leading edge. A similar situation exists in cladding where the molten pool leading edge is inclined at an angle and also when overlapping. These polarisation effects were not investigated in the present study.

The spatial power intensity distribution of a laser beam is not uniform. In fact, it can be a complex distribution depending on the beam mode. Two distribution patterns commonly found in gas lasers are the Gaussian and the 'dough-nut', Fig (13).

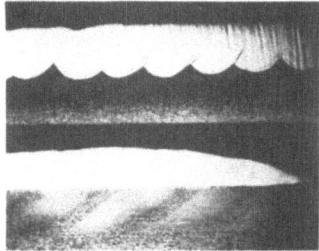

Fig. 12
Control of Dilution by powder feed rate.
top: powder feed rate was 0.09 g/s.
bottom: "   "   "   " 0.212 g/s"

Fig (13) A burn print produced
by a 'dough-nut' beam
mode.

The Gaussian beam is superior for cutting and welding
applications (ref 16) whereas for surface treatment the
'dough-nut' mode is preferred because of its more uniform
heating effect. In cladding this effect is utilised to
minimise dilution and obtain a desirable bead profile. Beam
symmetry is also important: Fig (14) illustrates the effect of
a 'detuned' mode structure with an off-centered peak.

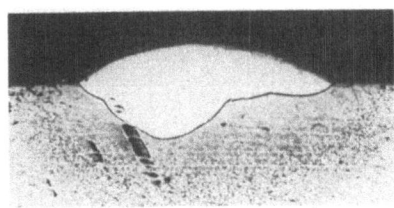

Fig. (14) Effect of a 'detuned'
mode with an off-
centered peak.

An inherent advantage of using a laser as a heat source is
that the source input can be localised to a small area.
However, such localised heating causes rapid self-quenching
and steep temperature gradients. In melting, these effects
often cause solidification cracking.

In the present study, a low power density i.e. a large spot
diameter of 5mm is used, coupled with slow speeds to produce
crack free clad layers.

METALLURGICAL AND PHYSICAL ASPECTS OF A LASER CLAD AUSTENITIC STAINLESS 316 LAYER.

A detailed evaluation (including corrosion properties) is the subject of a future paper. A summary is presented here in order to illustrate many aspects which may not be peculiar to stainless steel.

(a) <u>Surface Finish</u>. The surface finish of a clad layer produced by overlapping of single clad beads is characterised by the 'peak and valley' effect. (see Fig 2a). A typical 'peak to valley' distance of 40 microns can be obtained with a bead of acute contact angle (see Fig 11c) and a 60% overlap (as compared to 150 microns typical with arc weld overlay which needs to be removed to obtain a clear surface (after the plate is mangled flat). A typical value per unit area is $0.6Kg/m^2$.

b) <u>Porosity</u>. Porosity in laser clad layers may be caused by one or a combination of the following: cavities between two overlapped beads (usually formed near the root and will be referred to as inter-run porosity), solidification cavities and/or gas evolution.

Solidification cavities occur as porosity at or near the interface. If the clad layer has a significantly higher melting point than the substrate, the solidification front may finish at the interface thus causing shrinkage cavities (e.g. stainless steels - aluminium).

In conventional welding, porosity attributed to evolution by way of dissolved gases and oxidation reactions is well known and such phenomena are also applicable in laser cladding.

In cladding stainless steel to mild steel, use of a good shrouding technique and use of dried powder limit the problem to inter-run porosity.

The formation of inter-run porosity is due to a purely 'physical' effect i.e. the contact angle of a single bead or the leading contact angle of an overlapped bead, Fig (15a,b,c).

In general, a bead width/height ratio of greater than five (i.e. contact angle 43%) and a percentage overlap of less than 70% will produce clad layers of no inter-run porosity.

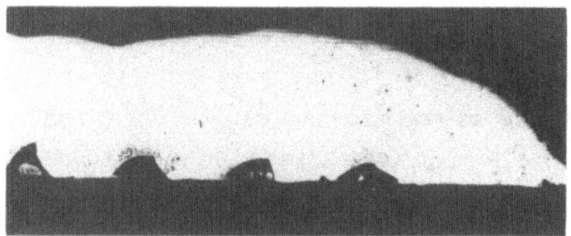

Fig (15a).

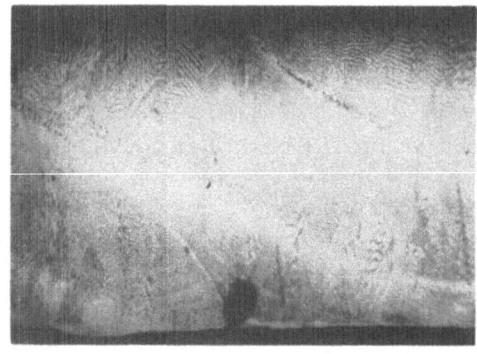

Fig (15b)

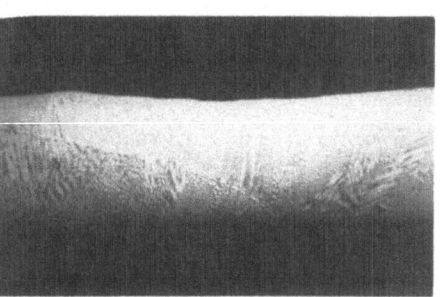

Fig (15c)

Fig (15) (a) Inter-run porosity. (b) Location of pore in relation to interfaces shown by etching. The subtrate appears totally blackened. (c) A clad layer with no porosity produced with beads of low contact angle.

(c) <u>Residual</u> <u>Stress/Plate</u> <u>Distortion</u>. Distortion of laser clad plates is very small1, Fig (16). There is minimal thermal penetration of the substrate, the heat affected zone being of the order 1-2mm.

Surface residual stresses were measured using the X-ray diffraction method. High tensile stresses of the order 285MPa were found on the clad surfaces. Using a pre-tensioning technique, it was possible to reduce the residual stress dramatically to compressive 8MPa (ref. 9).

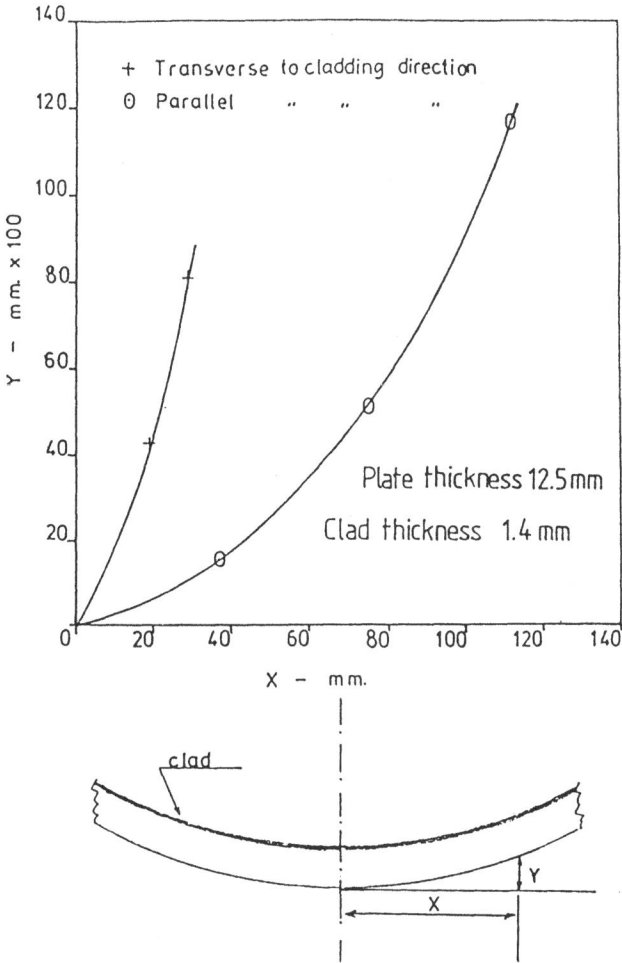

Fig (16) Distortion of a laser clad plate

(d) <u>Mechanical</u> <u>Properties</u> satisfactory results were obtained from tensile, bend and shear tests carried out in accordance to ASTM 264, Fig (17a.b).

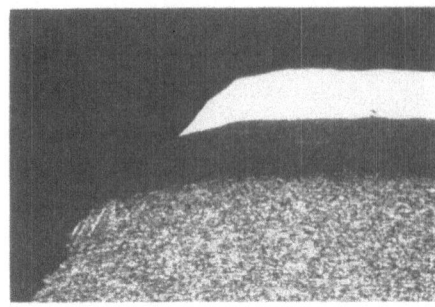

Fig (17a) Section of fracture surface of tensile specimen showing no delamination of the clad layer. Clad thickness 1.5mm.

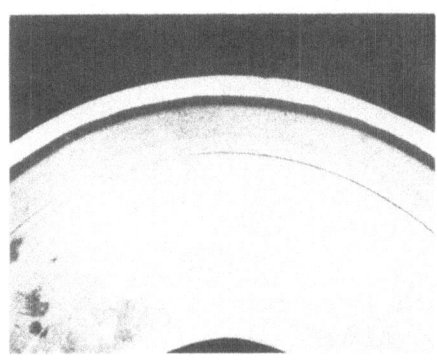

Fig (17b) Bend test specimen 12.5mm thick mild steel clad with 1.5mm thick stainless.

(e) <u>Microstructure</u> The dendritic microstructure of the laser clad stainless 316L layer was observed to be basically similar to that of a conventional weld bead or a submerged arc strip clad, i.e. austenitic matrix with residual delta ferrite. The dendrite size is much finer due to the higher cooling rate, Fig (18a,b).

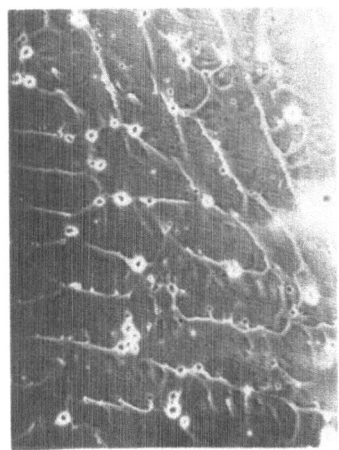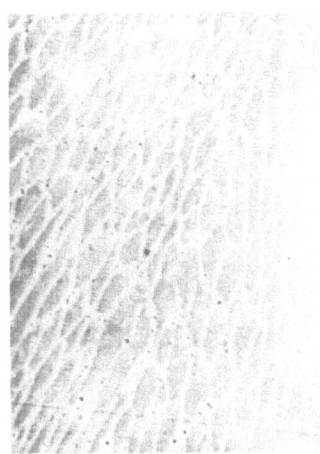

Fig(18)Effect of cooling rate on dendrite size - S.E.M.
microgrpahs.

    (a) Cladding speed 7mm/sec, 92 W/mm^2

        Dendrite size 6 microns

    (b) Cladding speed 40mm/sec, 573 W/mm^2

        Dendrite size 2 microns

(f) Cracking. In hardfacing, cracking is related to the
hardness of the clad layer and can be eliminated by substrate
preheat, fig (19). In austenitic stainless steels, cracking is
known to take place at high temperatures (1200°C) and as
such preheating has proved to be ineffective (Ref. 17).

Interdendritic cracking was observed in stainless clad layers
produced at high cladding speeds (20mm/s) with high power
densities (366W/mm^2). Crack free clad layers were produced
at lower power densities coupled with slower speeds.

Fig (19) Progressive elimination of cracking by pre-heating
          the substrate. Clad-Iron Boron (1000 V Hv)
          Substrate - Mild Steel.

(g) <u>Chemical Composition/Homogenity/Dilution</u>. Fig (20a,b),
illustrate the distribution of alloying elements in a 316
stainless clad layer and in the interface region, showing no
macro-segregation and negligible dilution.

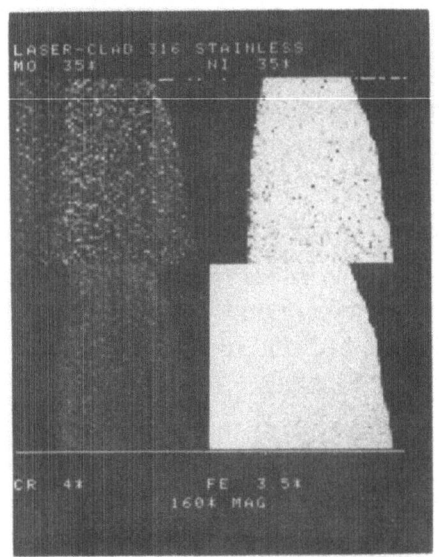

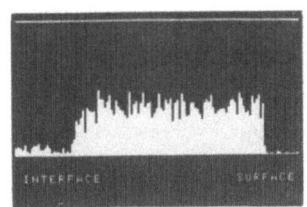

Fig 20b):
Line scan of chromium across
the clad/substrate interface.

Fig (20a) X-ray digimap of
alloying elements in a
316 stainless clad layer

One may define dilution as the compositional difference between the injected powder and the clad layer. Zero dilution was observed near the clad surface, with an overall dilution of about 5% typical. Also Fig.(21) shows that even with high penetration dilution is confined to the interfacial region, indicating perhaps the 'calmness' of the process.

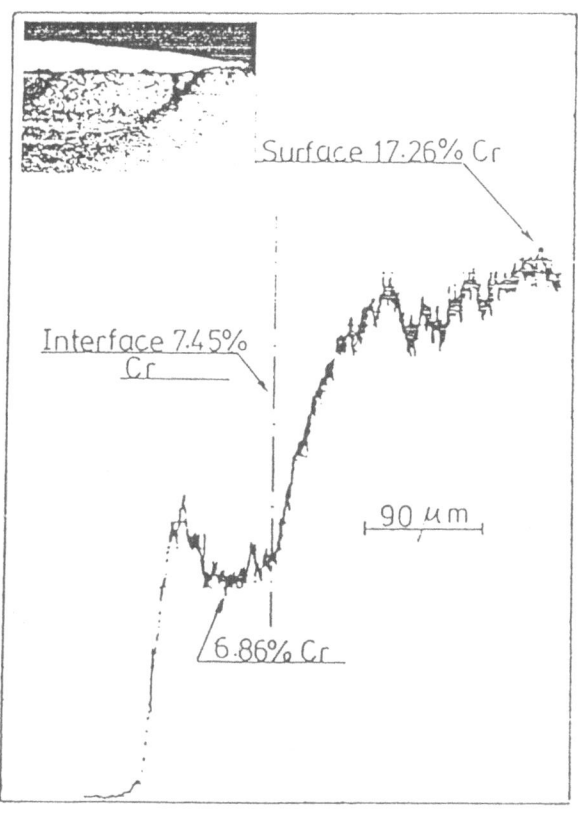

Fig (21) Crystal spectrometer scan of chromium across the thickness of a clad layer with high penetration.

OTHER CLAD/SUBSTRATE COMBINATIONS

In addition to stainless/mild steel, the following clad/substrate combinations were produced using the laser.

Note: Stellite SF6 a) Nickel/mildsteel
is a cobalt based b) Bronze/mildsteel   Some porosity but
hardfacing alloy. c) Stellite SF6/brass good bond strength.
d) Chromium/titanium
e) Stainless/aluminium Cracking plus dela-
f) Iron boron/mildsteel   mination in 'e'.
g) Stellite SF6/mildsteel
h) Mildsteel/Stainless

SOME SPECIAL APPLICATIONS.

Figs (22a,b) illustrate hardfacing applications. Figs (23a,b) show composite clad layers of hard and soft materials.

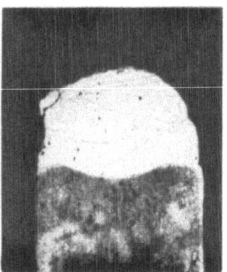

Fig (22b) Edge of a 2mm thick
turbineblade laser
hard-faced using
stellite 30mm/sec 15 x

Fig (22a) A drainage plough
blade laser hard-faced with
Stellite SF6. Coating thickness 1.5mm
Processing time 7/10 mins/unit. full size

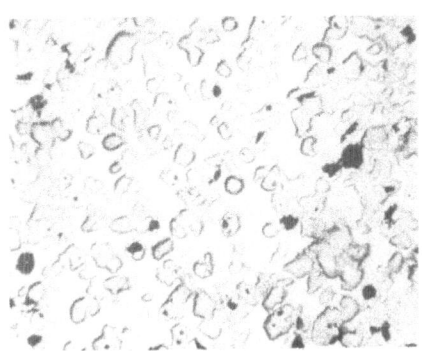 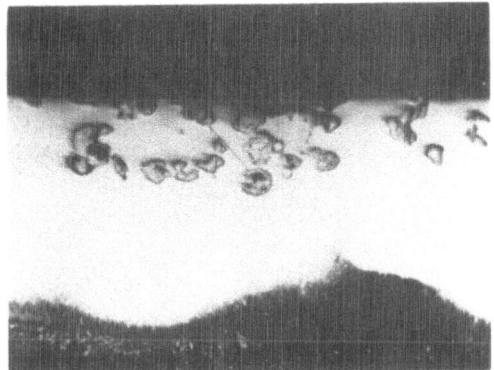

Fig.(23a) Titanium carbide particles in a stainless 316 clad matrix. Produced by pre-mixing the two powders. Carbide particle size 70 microns.

Fig (23b) Section of clad surface shown in Fig (23a). 50X.

CONCLUSIONS.

The process may be evaluated in terms of economic, engineering and quality assurance aspects.

Engineering aspects present less of a problem since they are mostly similar to other laser processing applications which have already found industrial acceptance; except perhaps aspects of powder injection, which have been developed to a reliable level during the course of the present study.

From a metallurgical/physical evaluation of a representative clad/substrate combination of austenitic stainless steel and mild steel, there is conclusive evidence that clad layers of acceptable quality can be produced using the laser.

There is no doubt, that laser cladding is best suited for treatment of small confined areas such as valve seatings, turbine blade edges, wear surfaces of tools etc. Deposition of thin coatings of expensive materials is also a good application: Laser cladding of large flat plates is envisaged to be carried out as an adaptive automated process, perhaps on a low priority in a multi-purpose time-sharing laser processing facility.

REFERENCES

1. F.G. Seaman, D.S.Gnanamuthu, 'Using the Industrial Laser to Surface Harden and Alloy'- Metal Progress, vol. 108, Nº 3 1975, p.67.

2. W.M. Steen, C.G.H. Courtney, 'Hardfacing of Nimonic 75 Using a 2KW Continuous Wave CO_2 Laser'. - Metals Techonology, June 1980. pp. 232-237.

3. W.M. Steen 'Surface Coating Using a Laser' Paper 3 Conf. Advances in Surface Coating Technology, Welding Int. London, Feb.1978

4. J. Powell, W.M. Steen 'Vibro Laser Cladding' Conf. Proceedings
    The Metallurgical Society of AIME, Feb 1981.

5. A. Belmondo, M. Castagna, 'Wear-resistant Coatings by Laser Processing'- Thin solid films, 64 (1979) p.249-256.

6. E.M. Breinan, B.H. Kear & C.M. Banas 'Processing Materials with Lasers 'Physics Today. Nov. 1976. p.44-50.

7. B.B. Moreton. 'Copper-Nickel Clad Steel for Marine Use' - The Metallurgist, May, 1976. p.247-252.

8. M. McIntyre, 'Laser Cladding at Rolls Royce' 2nd Int. Conf. on Applications of Lasers in Materials Processing, Jan. 1983, Los Angeles. USA

9. V.M. Weerasinghe, Ph. D Thesis 1985.Univ. of London.

10. U.K. Patent No. ...   , Quantum Laser Corp. , U.S.A.

11. J.Powell, Ph. D. Thesis 1983, Univ. of London.

12. V.M. Weerasinghe, W.M. Steen 'Computer Simulation Model for Laser Cladding by Powder Injection', Conf. Proc. ASME, Boston Nov. 1983.

13. G.G. Lim, W.M. Steen 'Measurement of Temporal and Spatial Power Distribution of a High Powered CO_2 Laser Beam' Optics and Laser Tech., June 1982.

14. A.K.L. Brochure, Hans Gressel weg. Munchen - 8000, W. Germany 1982.

15. J.N. Kamalu. Ph. D Thesis, 1981. Univ. of London.

16. J. Alexander, Ph. D. Thesis, 1982 Univ. of London.

17. R. Castro, J.J. Gadenet, 'Welding Metallurgy of Stainless and Heat Resisting Steels, Cambridge Univ. Press. 1974.

LASER METALWORKING

P. PIZZI
Fiat Research Centre, Strada Torino, 50, 10043 Orbassano (Torino), ITALY

1. LASERS METALWORKING APPLICATIONS : ADVANTAGES AND DISADVANTAGES

After more than 30 years from the invention of the laser the potentiality connected with deposition of high energy in materials has now become a well developing technology.

The laser tool in fact allowing to apply a very high energy flux of energy to the surface of materials, is concerning a very large number of manufacturing processes like cutting, drilling, welding, machining and surface hardening.

The lasers, electro-optical devices for converting electrical energy into a coherent beam of electromagnetic energy are fundamentally categorized in terms of the lasing medium, the shape of the laser beam-CW or pulsed, and the average power output of the beam.

One of the most important is the gas laser, the CO_2 laser; the others are solid state : ruby laser, neodymium doped ytrium-alluminum-garnet (Nd-YAG) laser.

The CO_2 and YAG are the most applied lasers in metalworking.

The CO_2 lasers operates at a 10.6 μm wavelength with a power output in the multikilowatt range, the YAG laser operates at a wavelength of 1.06 μm with an output power generally lower than 600 Watt.

The wavelength is very important because of the absorption characteristics of the materials.

The focused beam is directed to the workpiece to provide a sufficient energy to melt or heat treating the material depending on the processing, physical properties and surface characteristics of the material itself.

The high energy is absorbed in a outer layer about 10 nanometers thick; the high local power density of the beam reduces the induced thermal distortion, cracks or large stresses.

A second very important fact is connected with the absence of mass allowing to move and control the beam easily and in very short time.

From a general point of view the laser beam gives the following characteristics :
- acts at a distance in ambient atmosphere
- is applicable to any surface
- transfers very high energy fluxes to materials
- is a high flexible tool to perform different processes.

The principal disadvantages are concerning :
- high initial costs
- initial difficulty in maintenance, operation, installation in comparison to conventional processes.

Manufacturing systems are changing quite rapidly with the introduction of robotics, computer aided manufacturing and flexible manufacture.

Soares, O.D.D., Perez-Amor, M. (eds), Applied Laser Tooling. ISBN-13: 978-94-010-8096-5
© *1987. Martinus Nijhoff Publishers, Dordrecht.*

The general predictions for future manufacturing systems are that they will have [1] :
- high flexibility
- the ability to bring together workpieces and machining at the right time
- high productivity level
- highly qualified people to perform complicated tasks.

In this context laser technology represents an important tool whose development level can be summarized in Table I [1].

2. LASER - ASSISTED MACHINING
2.1. Introduction
Machining of materials with a cutting tool is one of the most common manufacturing activity in industry.

Currently, over 100 billion dollars per year are spent in the U.S. alone [2].

Machinability is generally characterized by the following parameters :
- tool wear
- cutting forces
- chip form
- surface quality.

These parameters are mainly influenced by the properties of the workpiece; heating of the workpiece to improve machinability influences the mechanical behaviour of the material reducing the cutting forces.

Reduction of the cutting forces and the possible increase in cutting speed leads directly to an intensive use of the manufacturing system. The cutting speed ranges of machining processes are shown in Fig. 2.1 [3].

The most important advantages induced in hot machining can be summarized as follows :
- cutting forces are greatly reduced
- power absorption and mechanical induced stresses are reduced
- tool life is increased.

Heating of the workpiece to improve machinability is a well known method arising from 1940 about [4].

Various approaches have been developed like :
- furnace
- flame
- electrical resistance
- arc heating
- plasma heating
- induction heating.

The "ideal source" requires local heating of chip formation avoiding influence of other areas where is essentially dangerous both for energy and local deformation criteria.

The laser source can be considered basically an ideal source in this context.

In fact this heating method generate a fast increase in the local temperature of the workpiece better than the others conventional sources.

2.2. Examples
To have an approximate understanding of the relationship between the cutting forces and workpiece temperature, in the turning of steel (38NCD4), a typical calculation for a cutting speed of 100 m/min is shown in Fig. 2.2 [5].

To define the optimal conditions for laser assisted machining [LAM], it is necessary to keep into account many parameters of the process like :
- spot dimensions
- power density
- beam position
- surface absorption.

During turning the workpiece moves in two ways with respect to the beam : a linear displacement (Vf, feed speed) and a rotation (V, cutting speed) giving, as a result, that the heated zone is moving along an helix.

However at a first approximation Vf<<V.

Considering, as an example, a material element running at a speed V under the laser beam during the interaction time, temperature is changing as shown in Fig. 2.3 [5].

The temperature evolution allows to deduce the optimal spot dimension and operating conditions.

In the case of the tests performed at Fiat Research Center for turning of various workmaterials like 38NCD4, 19NCD5 and 100C6 the cutting force reduction versus power density is shown in Fig. 2.4 [5].

It is to be remarked the large reduction of the cutting force only at low power densities; at values higher than about 0,3 kW/mm^2 the cutting force is quite constant and both the microstructure and the roughness of the material is modified.

Analogous experiences, performed in the DARPA, Advanced Machining Research Program, on titanium 6Al-4V with a 15 kW, continuous wave CO_2 laser have shown a two-fold increase in feed-rate with respect to the conventional machining.

In the case of ceramic tools turning Inconel 718 workpiece, LAM has given 40% less tool wear and 18% less cutting force than in conventional turning.

One of the most important parameter to be considered for economical reasons is the laser power; the usual absorption of a metallic surface is very poor and are in the order of 10÷15% of the incident power.

Generally, coating processes are used to improve absorption, like phosphating or sandblast. An example of absorption coefficients for different coating is given in Fig. 2.5 [6].

We can observe that for the same coating, absorption coefficients on carbon steel are 25% about higher than on nodular cast iron.

Enhanced absorptivity of metals and alloys for laser energy can be achieved by using a 1.06 μm wavelength by a Nd-YAG pulse laser.

The second point to be remarked is concerning the capital investment required for a pulse laser with respect to the continuous CO_2 laser.

Experiments performed on Inconel 728 and titanium 6Al-4V with pulse LAM have shown a cutting force reduction of 49% and 30% respectively [7].

2.3. Conclusions

The LAM, on the basis of the work performed on this subject, has some advantages with respect to others heating techniques :
- accuracy in the heat location source
- good control of operating parameters
- fast increase of the local temperature.

On the other hand the following problems have to be solved for LAM applications in workshops :

- cost reduction of the laser system
- safety problems for laser machining.

The economic analysis performed for LAM using a high-power, continuous wave CO_2 laser shows that the return of investment (ROI) is too low to justify the investments [7].

However, the use of low power Nd-YAG pulse LAM may be economically interesting for some specific applications.

The development of this technology is strictly connected with the development of new laser sources allowing a great cost reduction of capital investments.

3. LASER-ROBOTICS

Laser applications concerning cutting, welding, heat treating and alloying of materials are developed at different levels as shown in Table I.

If we keep into account the features of the laser it is evident the importance of the laser beam as a working tool; in fact the advantages of the laser processing can be summarized as shown in Table II.

As shown in Table II, flexibility, high power density and transmission of the laser beam are the most important features for materials processing.

While the use of high-power lasers for material processing is well known, the combination of high-power laser technology and robotics is to be developed. The essential element to provide the link between high power lasers and robots is the optical waveguide.

3.1. Optical waveguides

Depending on the type of high-power lasers, different optical waveguides can be used in material processing.

For example, in the near-infrared region of 1÷2 µm for Nd-YAG lasers, silica glass fibers have excellent transmission characteristics (Fig. 3.1).

The resulting schematic system using flexible optical fibers waveguide for delivering high-power laser radiation to the robotic arm/hand is shown in Fig. 3.2 [8].

Appropriate optical fiber waveguides for transmitting high-power laser radiation in the spectral region from UV to near IR allow to guide laser radiation from the source to the robotic arm.

For trasmitting high-power, longer infrared (2÷10 µm) a configuration similar to that shown in Fig. 3.2 can be used; however, presently long infrared fibers tend to be very lossy and fragile.

At present, hulky conventional articulating arms, consisting of aligned mirrors are used for applications requiring flexibility and transmission of high power infrared radiation.

The system configuration is strongly influenced by the motion of the laser beam with respect to the piece surface. In fact this situation will allow to design the laser-robotics system in function of the velocity and acceleration to which the means are subjected.

The class of optics used are basically two :
- refractive
- reflective.

The main advantages and disadvantages of the refractive and reflective optics are described in Table III.

Looking at Fig. 3.3, concerning the single laser source, the first trasparent optical system changing the beam characteristics and the operating heads, the following observations can be done :

- the configuration is useful for components to be moved relatively to the laser beam;
- the kinetics and dynamics are quite easy to be performed.

Generally, optical beams are submitted to multi-reflections to obtain high efficiency and flexibility of the laser beam (Fig. 3.4). These needs introduce different errors and particularly :
- errors in the focus head position
- errors in the laser beam mirrors reflection
- errors in gas-laser beam relative position.

The first error involve accuracy in the machining of the pieces, while the second is connected with the beam path and is concerning with mirrors accuracy and finally with beam quality.

The last type of error is particularly important in the laser cutting.

3.2. System configurations

The classification of the laser systems can be performed keeping into account the relative motion between the machining piece and the laser beam [9].

The two most important cases are concerning the cartesian robot and polar robot use as shown in Table IV.

In the first case it is achieved in general an higher accuracy and working space with respect to the polar robot.

However, the polar robot gives high flexibility and integration in the conventional factories.

Examples of the various systems are given as follows.

In Fig. 3.3 a schematic of stationary laser source single and multi-station with cartesian robot is given.

Systems of this type are commonly used in material processing, such as in the case of General Motors at Electromotive Diesel Division for cylinders line heat treating and those used at Fiat Auto Mirafiori Factory for synchro gears welding (Fig. 3.5).

This last application has been the first developed at Fiat Research Center in collaboration with COMAU and Fiat Auto SpA selected in function of the following needs [9]:
- applied to components usually produced in the automobile industry without changing the design;
- cost competitive with conventional processes
- quality increase.

A detail of the component is shown in Fig. 3.6.

The robot picks a gear from a pallet and places it on one position of a rotating table; after welding the gear is loaded on the pallet.

The synchronising gear laser weld is one of the most important laser applications in Fiat Auto Spa allowing to introduce correctly the laser technology in the factories to demonstrate the reliability and advantages.

In the cases where both for component size and for productivity reasons the laser beam is moved with respect to the piece it is necessary to design specific optics with increasing complexity.

Transeational and rotational axis are on the laser head as a working tool.

The supporting structure allows the movements of the "laser head" giving a systems configuration as shown schematically in Fig. 3.7.

The motions are generally divided between the "laser head" and the workpiece; in the case of no complex shapes as for sheet metal cutting the workpiece is generally moved with good accuracy but in other cases where the workpiece shape is very complex or large, it is suggested to use five axis on the "laser head".

An example of this type, for a five axis gantry system, to cut a deeply drawn sheet metal part is shown in Fig. 3.8. The system is to be installed at the Mirafiori Fiat Plant in Turin using a 5 kW laser source [9].

The "gantry type" configuration has generally certain disadvantages due to floor and volume, occupancy, rigidity, costs; on the other hand the automation in factories is well developed with the adoption of "polar robots" whose flexibility and costs are competitive with other systems.

The coupling of these polar robots with the laser is at the beginning of the development and generally four configuration can be distinguished :

   a. stationary laser source and fixed focal point;
   b. laser source robot handled (Fig. 3.9);
   c. laser source moving on one robot axis (Fig. 3.10);
   d. stationary laser source adapted to the polar robot by special optics (Figg. 3.11, 3.12).

The last solution improves flexibility of applications, accessibility and adoption of the conventional robots well used in production plants.

A mixed solution of a system constituted by a multi-head gantry with polar robots is shown in Fig. 3.13.

For the complexity of some processes like sheet metal cutting for car bodies the adoption of tracking sensors is very important and allows to compensate errors in the usual processing of materials.

The success of this "laser-robotics" technology is strictly connected with the integration of different technologies.

4. MEASUREMENTS OF OPERATING PARAMETERS

The laser for processing of materials is a very sensitive and flexible tool whose characteristics in shape, dimension, power density, intensity distribution are strictly connected with the quality of the process.

Considering that the process is the result of the laser-material interaction, the operating parameters for materials processing can be divided in two parts :

   1. direct laser beam parameters
   2. processing parameters.

4.1. Direct laser beam measurement

The direct laser beam measurement is a measure independent on the material processing and concern the laser beam intensity distribution in space and time. This type of measurement are difficult for high power beams in the infrared region, or expensive.

The power of the beam is measured usually by calorimetric techniques.

The calorimeters are based on measurement in steady state conditions of the thermal variations induced in a cooling fluid flowing in an absorbing target [10].

The block schemes of the calorimeter is shown in Figs. 4.1 a) and b) : the cooling fluid from the thermostat T is flowing to the laser radiation absorber by means of a pump P.

The laser absorber can be a cooled mirror or a sample support as shown in Fig. 4.2.

The flow rate f and temperature measurements allow to evaluate the power Q transferred to the cooling water, as follows :

$$Q = c \cdot f \cdot \Delta T$$

where : c is the specific heat and ΔT the thermal jump of the water.

The absorbed power can be measured directly or compared with a reference thermal jump electrical dissipator. A comparison between the obtained results are shown in Fig. 4.3 [10].

Beam shape and intensity distribution
We can summarize the methods for measuring the shape and intensity distribution of focused beams into two groups :

a. simple methods using vaporising materials where information is dependent on exposure time and is only comparative;

b. beam scanning with infrared or pyroelectric materials as shown in Fig. 4.4.

Fully or partially focused beams incident on the scanning head produce a time dependent voltage signal containing information about the beam intensity and shape. Electronics processing operate to display the signal [11].

4.2. Indirect measurements
During laser processing one of the most important parameter to be monitored to have a good control of the process is the surface temperature or emissivity.

A large improvement in this field is to be performed for future developments in laser technology.

An example of process control is given in the case of synchronizing ring welding to the transmission gear developed for Mirafiori Fiat plant.

Correlation between power and welding depth is shown in Fig. 4.5; these correlations allow to control the quality of the laser processing by monitoring the surface temperature at a specific point of the component. In fact surface temperature and welding depth are well correlated as shown in Fig. 4.6.

In the welding process many variables can influence the quality of the results as :
- rotational speed variations
- misalignment of the ring
- defocusing
- anomalous behaviour of the machine and optics.

These variations can influence surface temperature allowing to use a good process control [1].

REFERENCES

1. Pizzi P. : Lasers in Automobile Production. Nat. Symposium "Practical Application of Lasers in Manufacturing Industry". Coventry, 23 November 1982.

2. Komanduri R. : Tool Materials for Machining. Seminario su Lavorazioni d'Utensile ad Alta velocità. Torino, 3-5 June 1985.

3. Dieter K.A. : Inffentsammer. Seminario su Lavorazioni d'Utensile ad Alta Velocità. Torino, 3-5 June 1985.

4. Simonet J: Contribution to the Study of Hot Machining. Colloque Intern. sur la Coupe des Métaux, 21-23 November 1973.

5. Marinsek G., Capello G.: An Investigation on Laser Assisted Machining". Colloque Intern. sur la Coupe des Métaux, 21-23 Nov. 1973.

6. Gay P., Manassero G.: Absorption Measurements for High-Power Laser Measurements for High-power Laser Material Processing". ICALECO Conference, Vol. 38, 1983.

7. Komanduri R., Flom D.G., Lee M.: Highlights of the DARPA (Advanced Machining Research Program). Technical Inf. Series, Bldg 5 Room 321, Schenectady, N.Y. 12345.
8. High-power Laser and Optical Applications. The Bell System Technical Journal, No. 8, Part 1, Vol. 62, October 1983.
9. La Rocca A.V., Capello G., Pera L.: Some Considerations on the Definition of Laser-robots Systems. Conference "Laser Robotics I", Ann Arbor, Michigan, 22-24 April 1985.
10. Crescenzi F., Cutolo A., Gay P., Solimeno S.: Calorimetric Measurements of the Power Absorbed by Cooled Surfaces. Optical Engineering, Vol. 21, No. 3, May-June 1982.
11. Oakley P.J.: Measurement of Laser Performance. Practical Application of Lasers in Manufacturing Industry. Welding Institute, Coventry, 2-3 November 1982.

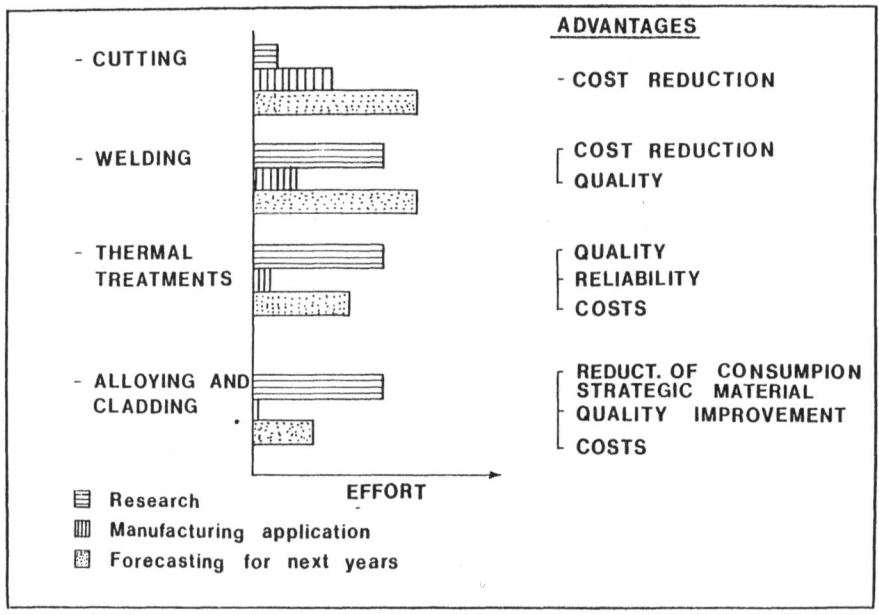

TABLE I. Research manufacturing and forecasted efforts for different laser applications.

PROCESS	BEAM POWER DENSITY /Watt cm^{-2}/	MAIN CHARACTERISTICS
CUTTING	$10^5 \div 10^8$	NO INERTIAL MASS, TO ENSURE ACCURACY AND FLEXIBILITY IN THE PROCESS
WELDING DRILLNG	$10^5 \div 10^8$	TRANSFERS OF ENERGY FLUXES OF EXTREMELY HIGH INTENSITY TO REDUCES DAMAGE IN THE WORK-PIECES AND DISPERSION
HARDENING	$10^3 \div 10^5$	FLEXIBILITY OF THE BEAM ALLOWS TO TREAT WORKPIECES OF COMPLEX SHAPES
GLAZING	$10^5 \div 10^8$	HIGH POWER DENSITY ALLOWS TO OBTAIN METALLURGICAL PROCESSES USUALLY DIFFICULT TO BE PERFORMED
ALLOYING	$10^3 \div 10^6$	FLEXIBILITY AND HIGH POWER DENSITIES ALLOW TO OBTAIN PARTICULAR METALLURGICAL PROCESSES IN MATERIALS

TABLE II. Laser processing characteristics.

TYPE	ADVANTAGES	DISADVANTAGES
REFRACTIVE	- TRANSPARENT - LOW MASS - CAN SEPARATE DIFFERENT VOLUMES	- FRAGILE - LOCAL ENVIROMENTAL SENSITIVITY - TRANSPARENCY AND REFRACTION DEPENDING ON WAVELENGTH - ANTI-REFLECTION COATINGS - THERMALLY SENSITIVE
REFLECTIVE	- RUGGED - EASY TO COOL AND PROTECT - OPTICAL PROPERTIES INDEPENDENT ON WAVELENGTH - EASY TO CLEAN - HANDLE HIGH POWER DENSITIES	- MORE MASSIVE

TABLE III. Advantages and disadvantages of laser optics.

1. **CARTESIAN ROBOT**   A.   STATIONARY LASER SOURCE

   - SINGLE STATION ⎫ OPTICAL HEAD + WORKING PIECE
   - MULTI STATION ⎬ MOVEMENTS

2. **POLAR ROBOT**   A.   STATIONARY LASER SOURCE AND FIXED FOCAL POINT

       B.   LASER SOURCE ROBOT HANDLED

       C.   LASER SOURCE MOVING ON ONE ROBOT AXIS

       D.   STATIONARY LASER SOURCE ADAPTED TO POLAR ROBOT BY SPECIAL OPTICS

TABLE IV. Cartesian and polar robots.

222

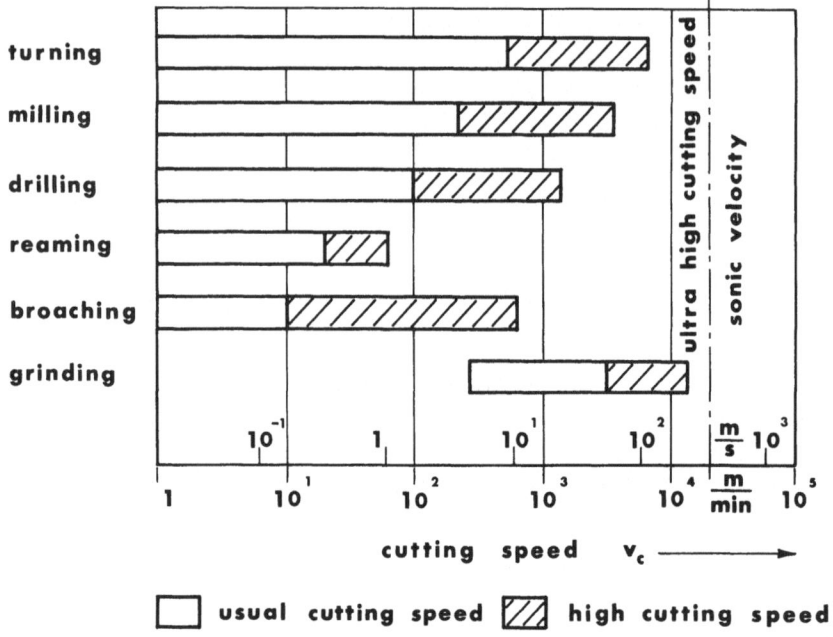

FIGURE 2.1. Cutting speed ranges of machining processes.

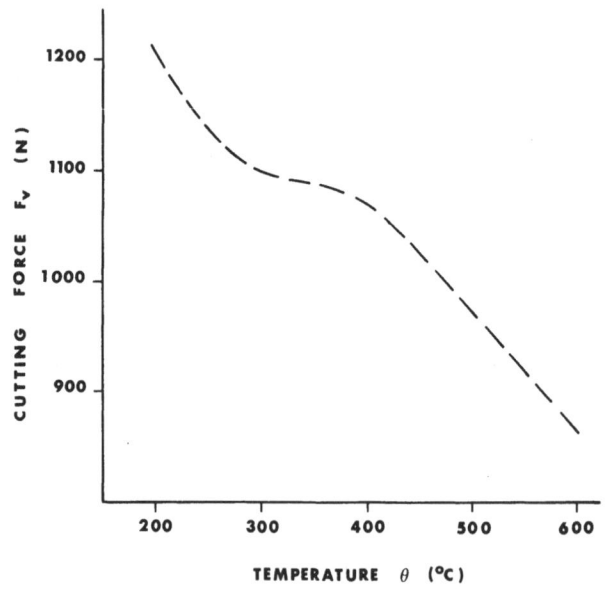

FIGURE 2.2. Correlation bewteen cutting forces and workpiece temperature.

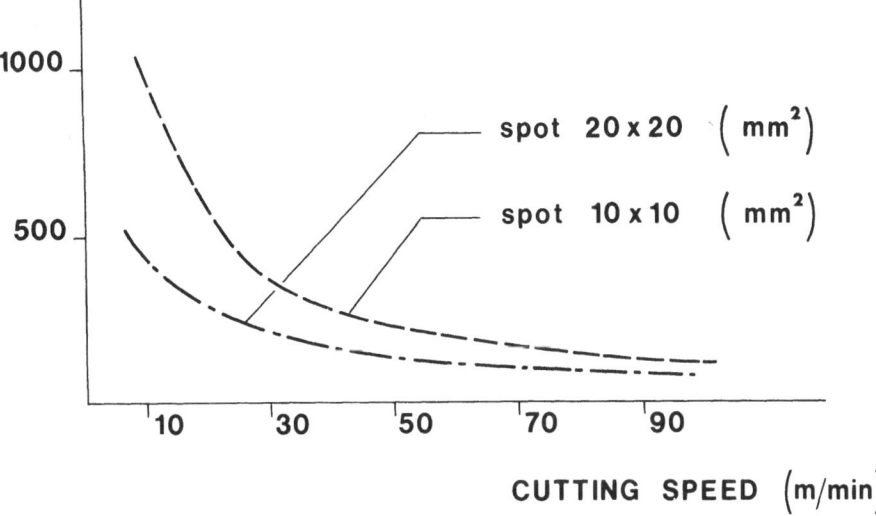

FIGURE 2.3. Temperature versus cutting speed with different spot size.

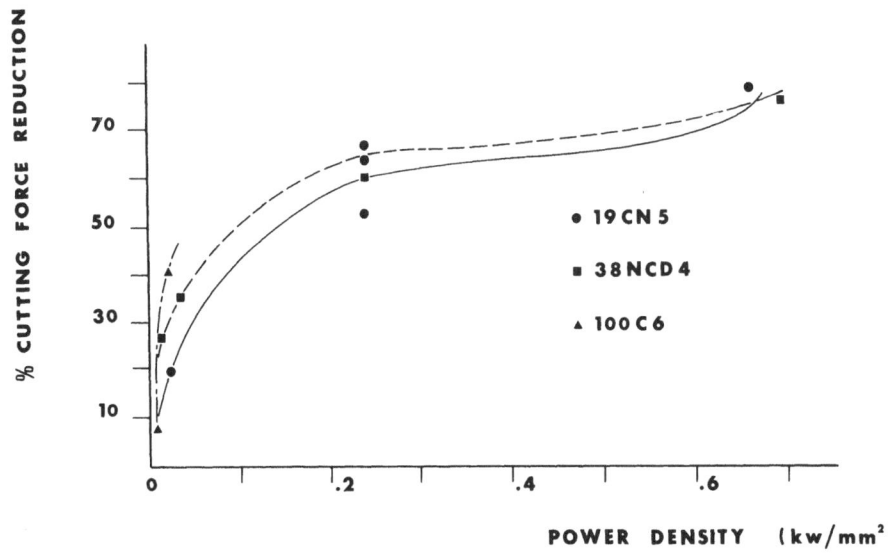

FIGURE 2.4. Correlation between cutting force reduction and power density.

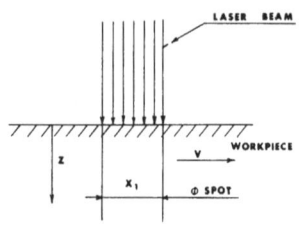

MATERIAL	SURFACE TREATMENT	AVERAGE INCIDENT POWER/W/	ABSORPTION COEFFIC. /A/AVERAGE	MAXIMUM TEMPERATURE /°C/ AT 1mm BELOW HEATED SURFACE
NODULAR CAST IRON	PHOSPHATE	964	0.59±14%	486
"	GRAPHITE	916	0.64± 4%	479
	SANDBLAST	952	0.33±18%	265
	TITANIUM OXIDE	911	0.71± 4%	564
AISI 1045	PHOSPHATE	870	0.89± 9%	436
	GRAPHITE	879	0.81±23%	396
	SANDBLAST	932	0.46±18%	218

FIGURE 2.5. Absorption coefficient for different surface coatings.

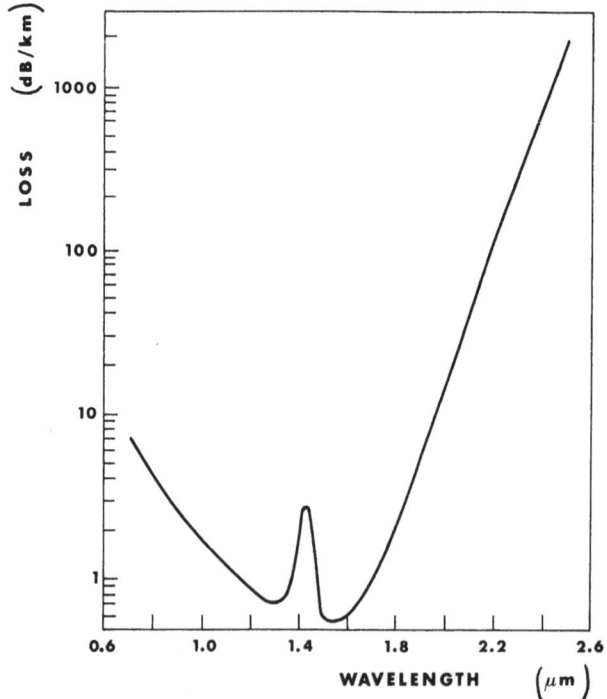

FIGURE 3.1. Spectra of typical low-loss silica glass fiber waveguide.

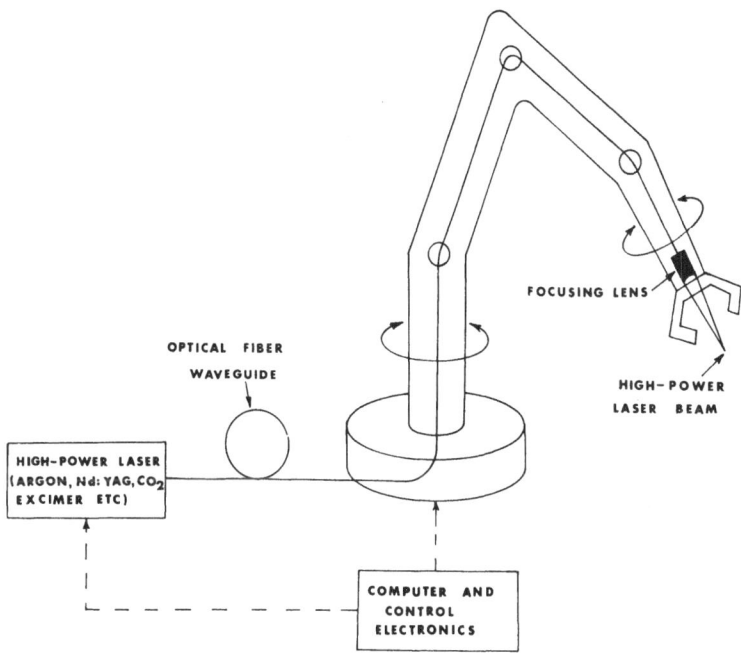

FIGURE 3.2. System using flexible optical fiber waveguide.

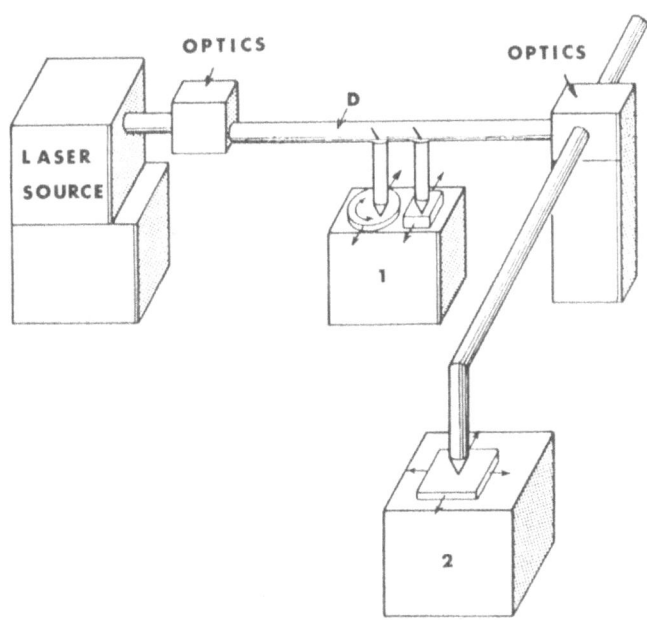

FIGURE 3.3. Schematic of stationary laser source multi-station with cartesian robot.

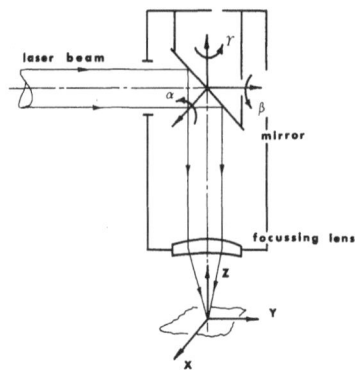

FIGURE 3.4. Schematic diagram for an optical focusing head.

FIGURE 3.6. Laser welded synchronising gear.

FIGURE 3.5. Two station robot assisted for synchro-gears welding in operation at Fiat Mirafiori, Turin.

227

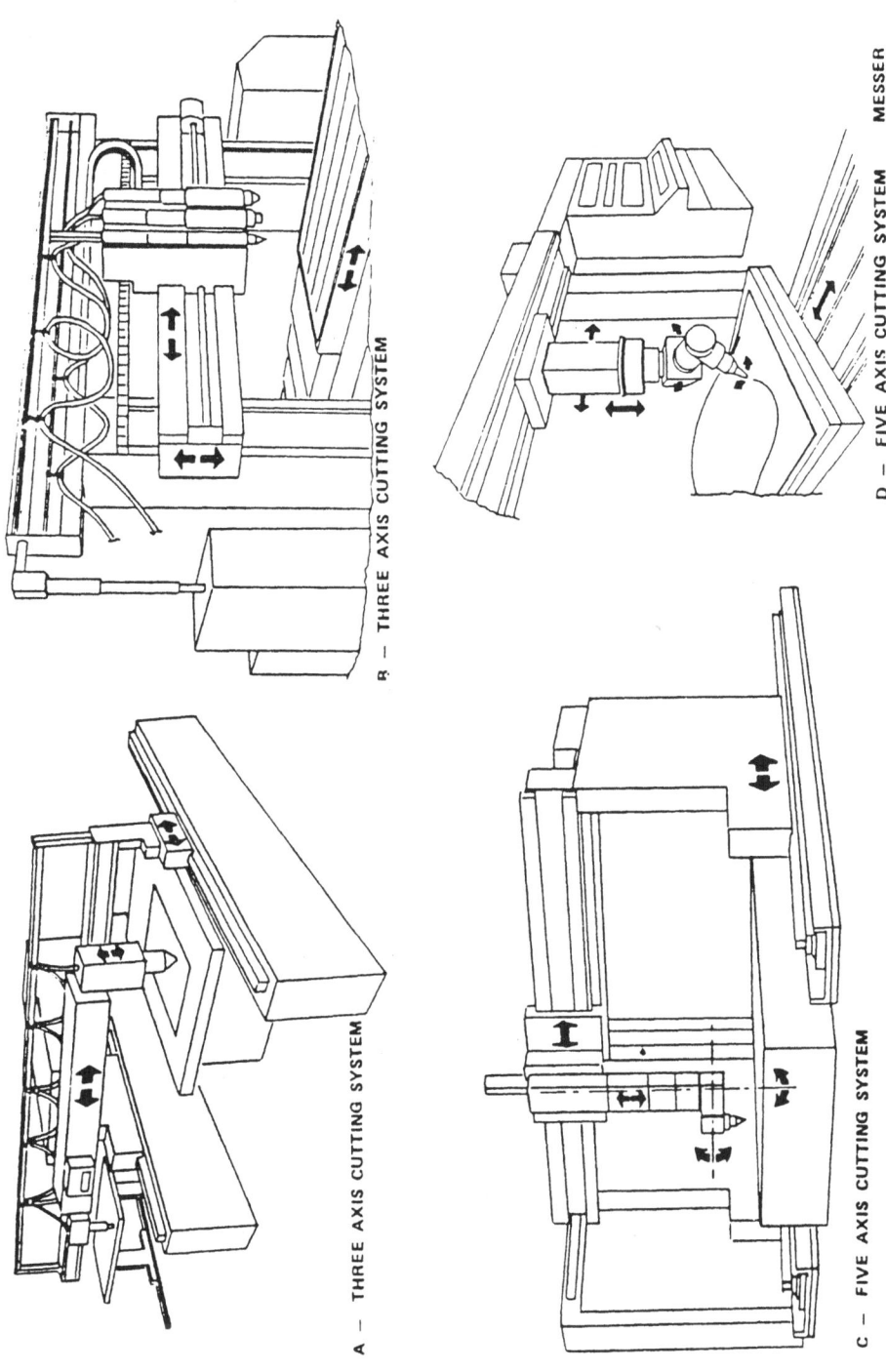

A — THREE AXIS CUTTING SYSTEM

B — THREE AXIS CUTTING SYSTEM

C — FIVE AXIS CUTTING SYSTEM

D — FIVE AXIS CUTTING SYSTEM   MESSER

FIGURE 3.7. Gantry system configurations.

228

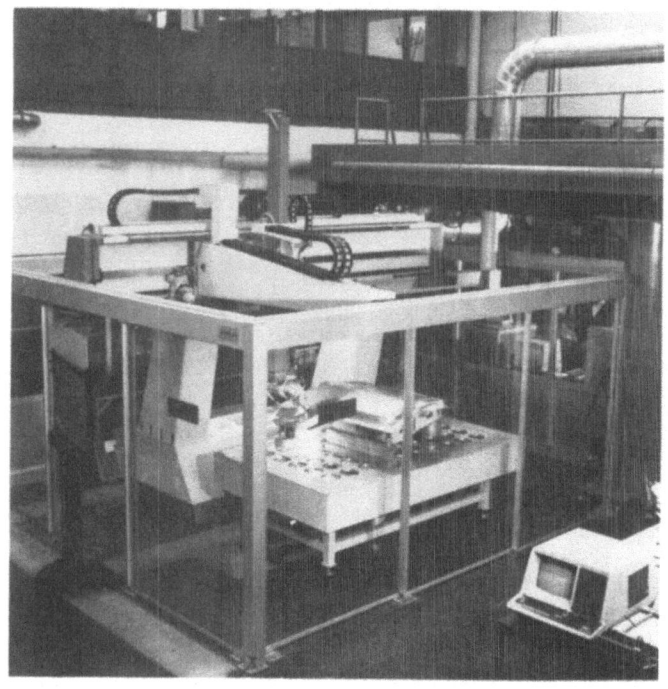

FIGURE 3.8. Gantry system with five motions axis on the head.

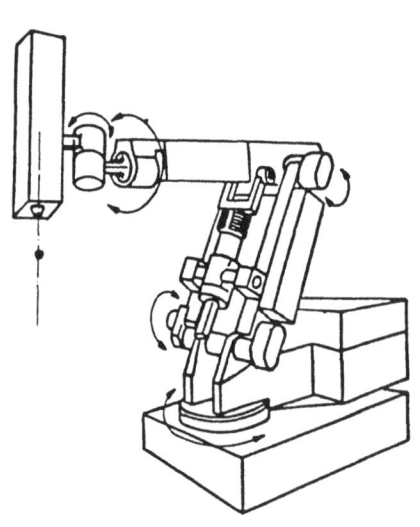

FIGURE 3.9. Laser source robot handled.

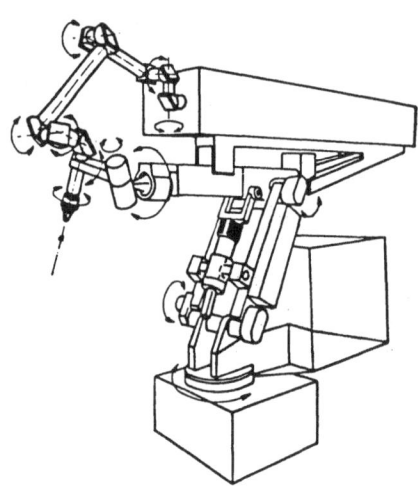

FIGURE 3.10. Laser source moving on one robot axis.

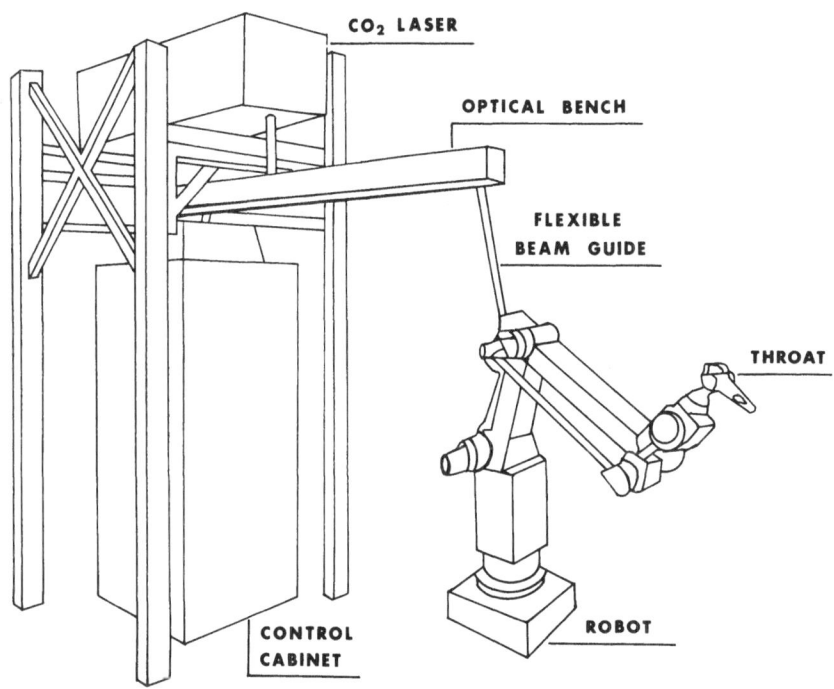

FIGURE 3.11. Stationary source adapted to the polar robot.

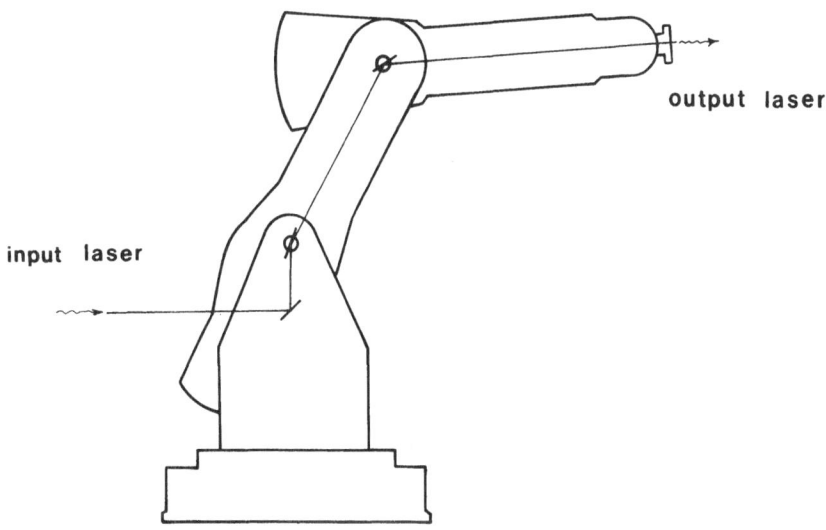

FIGURE 3.12. Schematic diagram showing the master configuration for CO_2 laser-robot system.

230

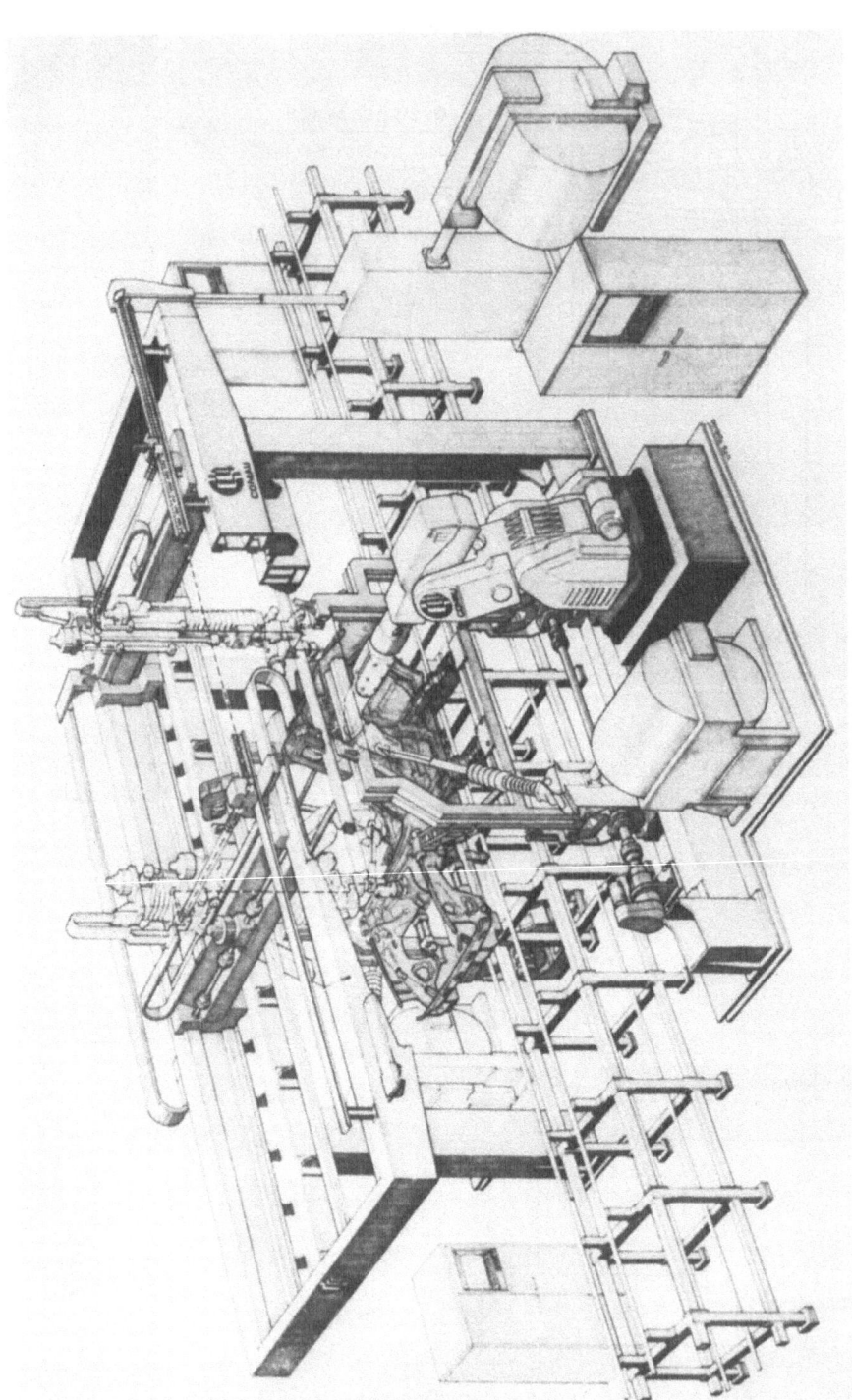

FIGURE 3.13. Multi-head gantry with polar robots.

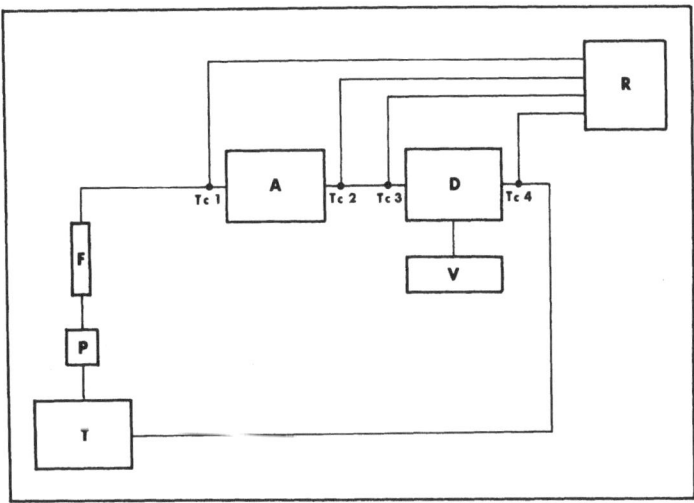

T = Thermostat
P = Pump
F = Flowmeter

A = Laser power absorbing target
D = Electrical dissipator
V = Recorder
T_{C1}, T_{C2}, T_{C3}, T_{C4} = Thermocouples

a) Flow scheme.

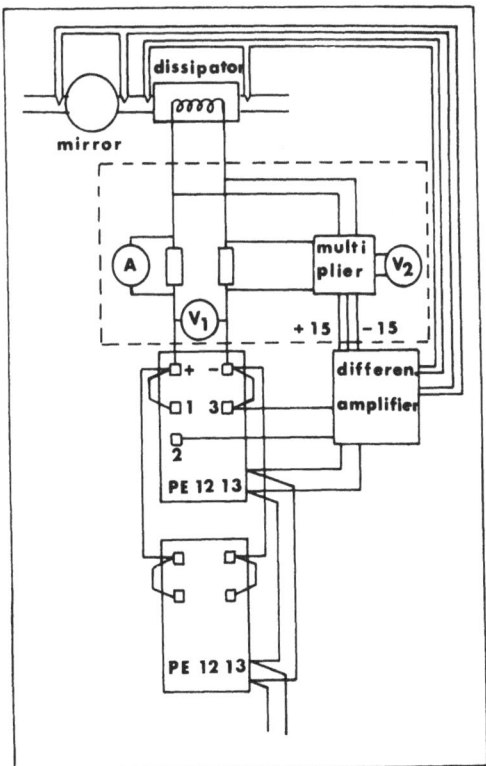

b) Electric block diagram.
The components PE 12 13 are stabilized voltage controlled power supplies.

FIGURE 4.1. Block schemes of the calorimeter.

232

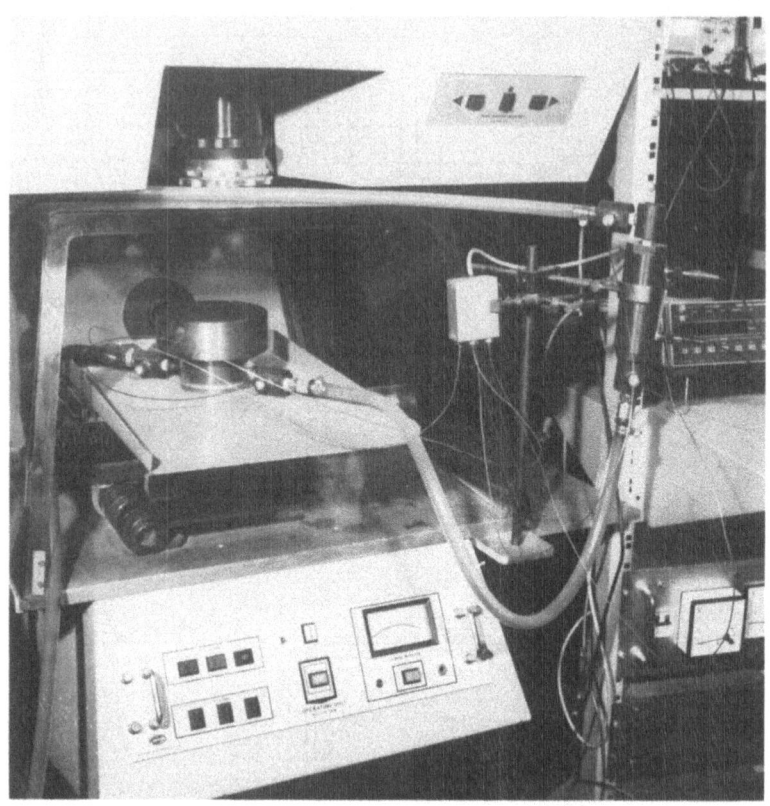

FIGURE 4.2. Photograph of a water cooled mirror during measurement of absorption with the calorimeter.

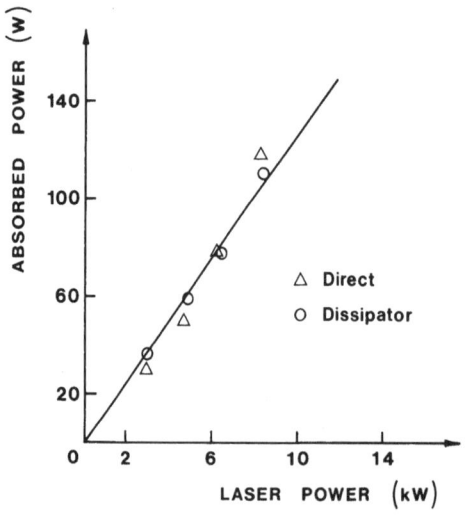

FIGURE 4.3. Power absorbed by a copper mirror versus the incident power.

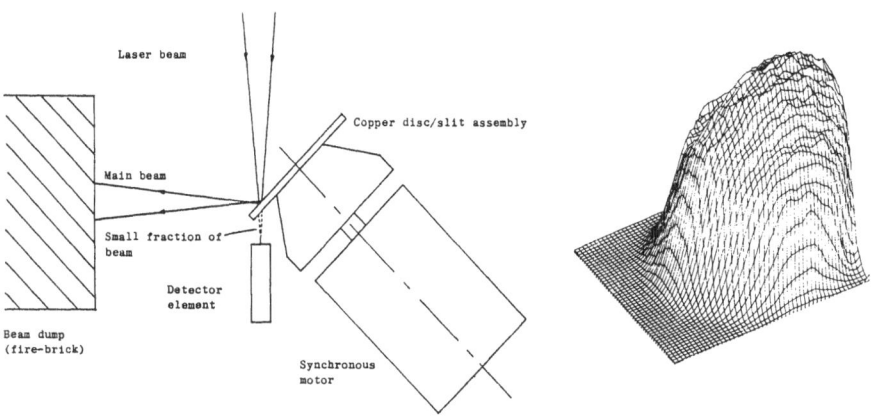

FIGURE 4.4. Laser beam scanner and intensity distribution.

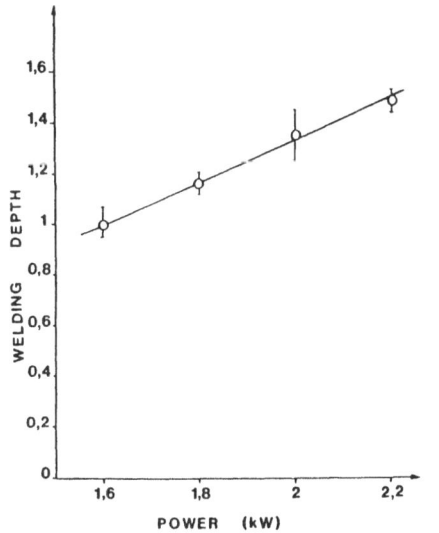

FIGURE 4.5. Correlation between welding depth and laser power.

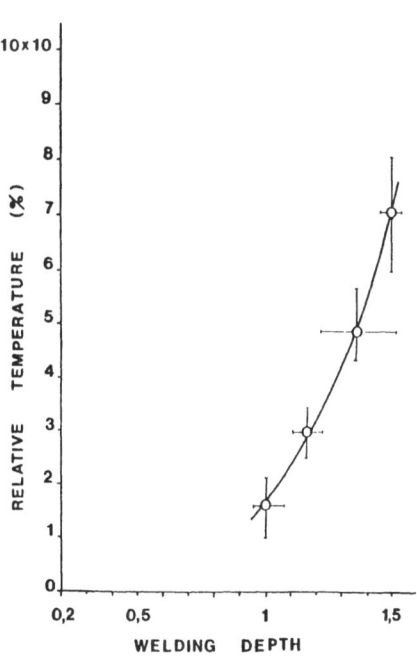

FIGURE 4.6. Correlation between welding depth and surface temperature.

LASER PROCESSING SYSTEMS IN PRODUCTION PLANTS

G.H. PARSONS
CONTROL LASER LIMITED, Daventry, Northamptonshire NN11 4NR, U.K.

1. INTRODUCTION

During the 1970's laser manufacturers were predominantly concerned with selling their basic laser generator and generally showed little interest in developing complete processing systems. The prospective user had the choice of either developing a system "in house" or involving a conventional machine tool or systems designer to buy a laser and incorporate it into a total system.

These options often proved unsatisfactory with the designer failing to understand the fundamentals of laser processing and producing either an unreliable system or perhaps an over-engineered solution.

Today, although many laser manufacturers have maintained this marketing approach, other more progressive companies have responded to the customer's needs, and are able to provide a standard or tailor made system, to perform the required process. The advantages of buying from a single supplier who understands, and is responsible for, the whole system are very obvious.

2. SYSTEM DESIGN

Laser processing implies relative movement between the beam and the work-piece which may be achieved by moving one or more of the three components laser, laser beam and workpiece, or sometimes by a combination of two of these.

2.1. Moving Laser

The workpiece is stationary and the laser is mounted on a single or two axis machine. This configuration is limited to small low power lasers (say less than 400W), and suffers from problems of high inertia which requires heavy and expensive machinery to produce a workable system.

2.2. Moving Workpiece

This technique is an obvious solution where the required movements are not too large or complex, involving one or two degrees of freedom. A good example would be two dimensional profiling using a CNC two axis table under a stationary beam.

In general, moving workpiece configurations requires a relatively light workpiece and are perhaps most valid where movements are limited to a maximum of 1m x 1m. Although larger tables are produced, they can be very expensive.

Attempts to achieve 3D workpiece manipulation using an industrial robot have not proved very successful.

2.3. Moving Optics

The laser beam is manipulated by optical techniques around a stationary workpiece. The method is probably the most elegant but requires both engineering and optics expertise to build a successful system. Machines can range from single to multi-axis movements and are particularly relevant where the need is for a flexible system capable of treating small parts or

Soares, O.D.D., Perez-Amor, M. (eds), Applied Laser Tooling. ISBN-13: 978-94-010-8096-5
© *1987. Martinus Nijhoff Publishers, Dordrecht.*

heavy components.

The moving optics approach has formed the basis for the evolution of systems design within Control Laser Limited. This paper describes four approaches which illustrate this development from simple beam switching to full 5-axis robotic manipulation.

3. BEAM SWITCHING

Laser cutting or 86mm diameter "Kevlar" tubing - The system was designed,Fig1. to perform two laser cutting operations:

Workstation A   Cut to length;

Workstation B   Produce a 30° chamfered cut at one end of the tube.

The process is based on a 500W F A F CO_2 laser with a switching mirror to switch the beam between the vertical cut nozzle of workstation A and the angled cut nozzle of workstation B.

The two workstations are part of an automated machining unit which includes hopper feed, conveyors, an automated lathe station and work-handling for the laser stations, Fig 2.

Cut length and quality are important and the unit includes automated inspection devices and is controlled by a sequence controller which is integrated with the laser controls.

The entire machining unit was designed and specified by Control Laser Limited and forms part of an automated factory.

4. SINGLE AXIS MOVING OPTICS

Laser cutting and welding of Silicon Steel - This system is installed in a Swedish steel rolling mill and its purpose is to join two 7 ton rolls of steel to produce a 14 ton roll for subsequent re-rolling to finish thickness, Fig 3.

The system is based on a 2kW F A F CO_2 laser which is optically linked to a single axis carriage mounted on a rigid stress relieved steel box section girder. The carriage moves on "Dexter" recirculating bearings and is driven by a servo motor and CNC controller. The carriage carries twin nozzles which are selected by a switching mirror. The first nozzle assembly is fitted with a 150mm focal length lens and is used for the cutting process, while the second nozzle, fitted with a 75mm focal length lens, performs the welding process.

The process sequence involves cutting the roll end to remove scrap and provide a true edge for the subsequent welding process. After the second cut the process controller activates the system hydraulics to bring the clamped edges into close contact, the beam is switched to the welding nozzle and the X carriage returns across the roll width welding the two rolls together.

The laser process replaces conventional TIG welding and mechanical shearing and has achieved the following benefits.

a) The width of the weld zone after rolling is reduced from typically 300mm for TIG to 30mm for laser.

b) The failure rate of welds during rolling has reduced from 60% for TIG to 10% for laser welds.

c) The life of the steel rolls has improved dramatically.

5. TWO AXIS MOVING OPTICS

The extension of moving optics into two axis machines was a logical step, building on the experience of single axis work using the basic structures of stress relieved box section frames, carrying Dexter recirculating bearings.

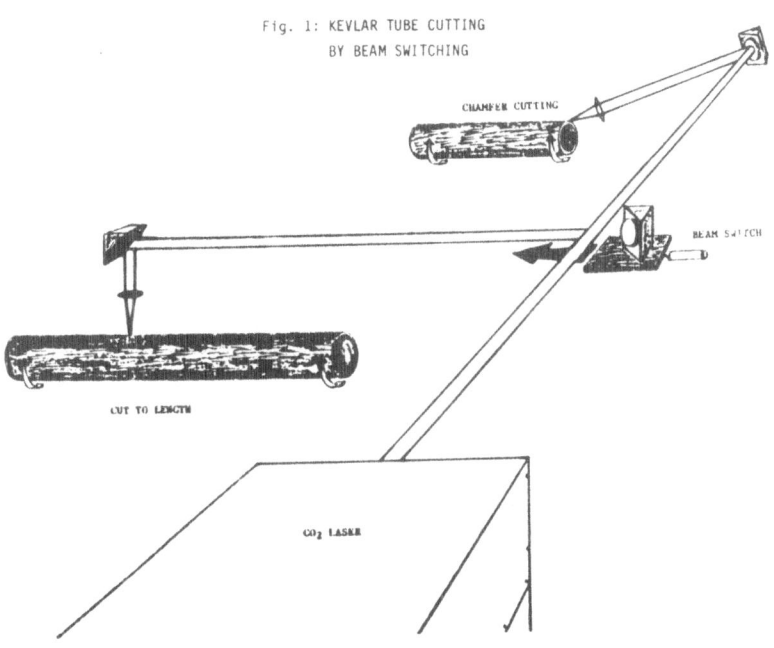

Fig. 1: KEVLAR TUBE CUTTING
BY BEAM SWITCHING

CHAMFER CUTTING

BEAM SWITCH

CUT TO LENGTH

CO$_2$ LASER

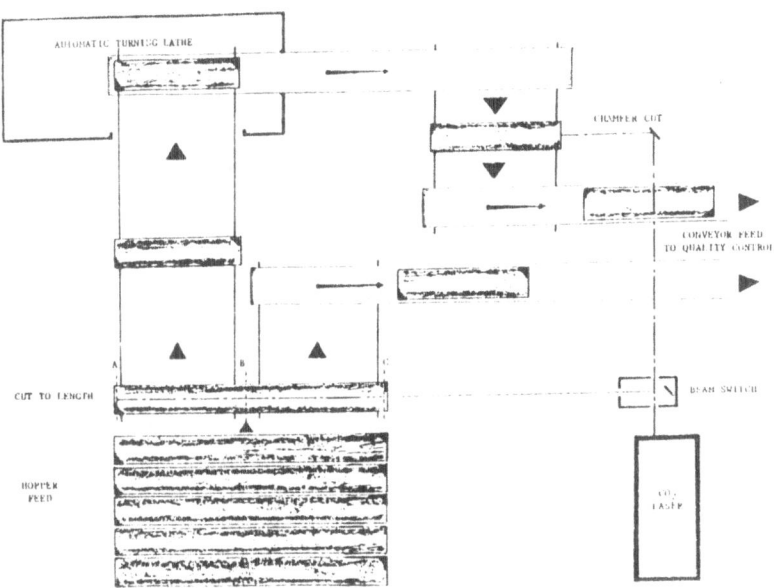

Fig. 2: PLAN VIEW OF MACHINING CENTRE

AUTOMATIC TURNING LATHE

CHAMFER CUT

CONVEYOR FEED
TO QUALITY CONTROL

CUT TO LENGTH

BEAM SWITCH

CO$_2$
LASER

HOPPER
FEED

238

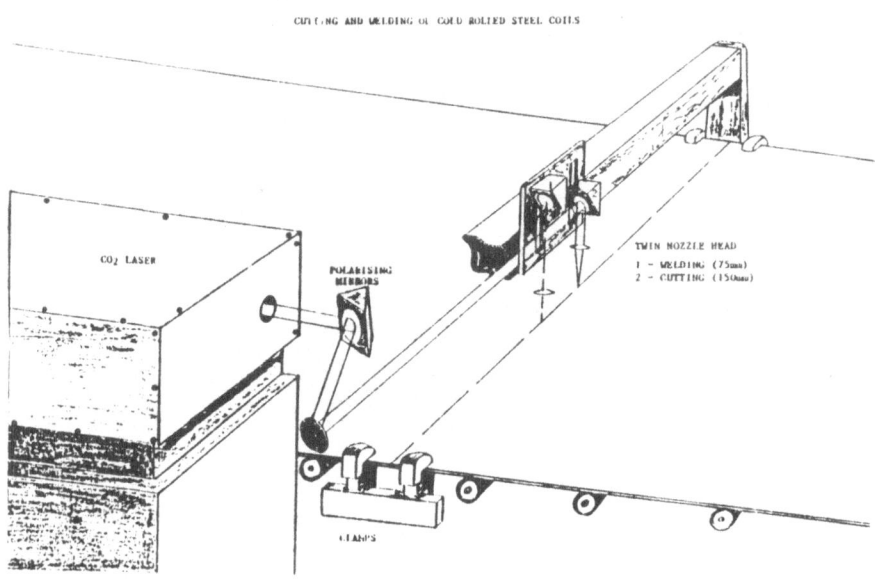

Fig. 3: SINGLE AXIS MOVING OPTIC

CUTTING AND WELDING OF COLD ROLLED STEEL COILS

CO₂ LASER

POLARISING MIRRORS

TWIN NOZZLE HEAD
1 - WELDING (75mm)
2 - CUTTING (150mm)

CLAMPS

Fig. 4: A MULTI AXIS MOVING OPTICS SYSTEM

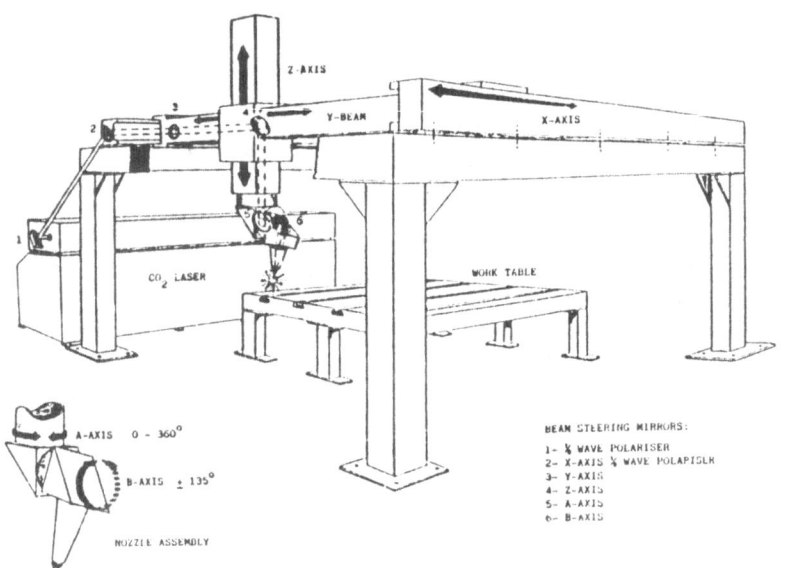

Z-AXIS

Y-BEAM

X-AXIS

CO₂ LASER

WORK TABLE

A-AXIS   0 - 360°

B-AXIS   ± 135°

NOZZLE ASSEMBLY

BEAM STEERING MIRRORS:
1- ½ WAVE POLARISER
2- X-AXIS ½ WAVE POLARISER
3- Y-AXIS
4- Z-AXIS
5- A-AXIS
6- B-AXIS

For a 2 axis machine a Y-axis carriage is mounted on a cross-beam which is itself running on twin X-axis frames. In a standard machine the Y-carriage carries a ride-on nozzle assembly which is pneumatically operated.

The laser beam optics are set in rectilinear coordinates and comprise 2 circular polariser mirrors in a link arm which drops the laser beam down to machine height, a collecting mirror at the end of the cross-beam and a Y-carriage mirror to direct the beam down via the lens and nozzle to the workpiece. The laser beam is totally enclosed from laser head to cutting nozzle within the frame of the machine.

Profiling control is by a Bosch Alpha 3 CNC driving the machine through DC Servo motors and linear encoders.

The 2 axis moving optics table is now a standard laser profiling system with a working area of 2½ x 1½m while other sizes can be designed to suit a users particular needs.

Systems in production plants cover a range of applications which include laser profiling jobbing shops, armour plate profiling, display sign manufacturing in metal and plastic, nuclear research and recently circular sawblade profiling.

6. MULTI AXIS MOVING OPTICS

Laser profiling of 3 dimensional shapes implies a significant technology leap compared with 2 dimensional work. In addition to movements in the Z plane, the machine must also make positional adjustments of the nozzle which involves at least 2 further movements usually termed A and B rotations, Fig 4.

6.1. Case Study Austin Rover Group

Background: In 1980 B.O.C.'s Industrial Power Beams subsidiary was bought by the Control Laser Corporation of Orlando, U.S.A. and became Control Laser Limited. The CO_2 laser technology which is still based in the U.K. company was augmented by technology interchange with CLC particularly in the field of robotics and control systems.

Against this background, discussions began in 1982/83 with Austin Rover Group (formerly British Leyland) on the feasibility of a 5 axis laser robot for trimming pre-production pressed panels.

6.1.1. Justification: Austin Rover identified the following benefits to be derived from a flexible laser trimming system.
1) Avoidance of hand cutting;
2) Ease of trim line adjustment;
3) Avoidance of trim die adjustment:
4) Earlier production of correct trim dies;
5) Models earlier into production.

The company was spending some £250,000 annually on trim die adjustment and rather more on hand trimming.

Based on these considerations the company decided to proceed and a machine that used the best of the USA's robotics experience with the UK company's wide experience of XY moving optics was designed, built and put into service in some 6 months.

6.1.2. The Machine: The machine consists of four large braced pillars which support X axis steel slideways. A twin cross-beam carries a Y-axis carriage, and suspended from this carriage between the twin beams is a Z axis slideway.

All three axis run on "Dexter" recirculating bearings and feature adjustable slideway mounts. The Y-axis cross-beam has double ended rack drives and separate linear encoders for controlled movement in the X-direction.

The Z-axis slide is based on an aluminium casting with pneumatic counter-balancing and a brake motor. This Z-axis slide carries the cutting nozzle assembly which has two further degrees of freedom comprising 365°, A-axis horizontal rotation and 90° B-axis vertical rotation.

6.1.3. Beam Delivery Optics: The large diameter, low divergence laser beam is generated by a 1.2kW CO_2 fast axial flow laser positioned along side the robot.

The laser beam is led to the cutting nozzle by 2 silicon ⅛ wave phase retarding mirrors, 4 silicon mirrors and a focussing lens. Beam alignment aids are built into the machine and maintenance of beam alignment is enhanced by positively located mirror mounts.

6.1.4. Control: Machine control is by a Laser Brain CNC unit developed by Control Laser Corporation. This controller features two microprocessors and computes robotic movements necessary to maintain position of nozzle tip in space during A and B axis rotations. The controller has an update time of 15 millisecs and a positional accuracy within the 2m x 2m x 0.75m working envelope of +/-0.15mm over 1m movement.

A wide range of options exist for data input including:
1) Automatice optical line following;
2) Teach in by manual positioning using an umbilicate pendant;
3) Manual keyboard;
4) Digitised input;
5) Modem link to C.A.D.

6.1.5. Safety: The safety requirements for a 5 axis laser machine are significantly more demanding than for a 2 axis system in which the beam direction is fixed, usually vertically downwards.

Control Laser engineers worked in close co-operation with the customer's safety officers and arrived at a safety package which provided additional to the standard laser safety provision, the following features:
1) A complete machine enclosure with safety doors interlocked with the beam operation;
2) weight detecting safety mats within the enclosure;
3) safety "kick-bars";
4) a special circuit which cut off the laser beam if the nozzle assembly ceased moving for more than five seconds. This 5 seconds period was arrived at following exhaustive tests on enclosure and window materials;
5) a pressure sensing device at the nozzle which prevents use of the beam in the absence of a focussing lens.

6.1.6. The Production Experience: It had been understood from the beginning of the project, by customer and manufacturer, that a period of learning and development was inevitable and that modifications and changes based on production experience might be necessary.

Plusses:-
1) The machine paid for itself in its first few months.
2) Operators learned techniques rapidly especially for manual teach-in.
3) Availability was in excess of 95% and the laser ran 6000 hours in the first fifteen months.
4) Maintenance record was good.
5) The basic robust engineering proved reliable.

Minusses:-
1) The A and B axis which were based on standard rotary tables proved to be too bulky and vulnerable to damage. These axes have been redesigned.
2) The automatic line follower proved hard to use on sharply profiled panels.
3) Metal particles shorted out cubicle safety mats.

The Unexpected:-

1) Enthusiastic reaction from initially sceptical shop floor management and operators.

2) Flexibility allowed utilisation in new areas such as low volume variant production and competition car cutting and welding.

3) Press tool hardening trials proved promising.

In summary the machine was a success because the customer involved the right people at the beginning. Any machine will work better if the workforce want it to!

APPLICATIONS OF LASERS IN PLANAR AND THREE-DIMENSIONAL SILICON INTEGRATED CIRCUIT FABRICATION

D.P. VU

CNET - CNS - Chemin du Vieux Chêne - B.P. : 98
F-38243 Meylan Cédex

Lasers as directed energy sources are finding more and more applications in microelectronics. This paper describes some of the most important results in this field showing the great potential of lasers for future development of Si integrated circuit technology. The paper is divided in 3 sections. We shall first discuss the use of lasers in device processing as an alternative to conventional furnaces for improving the performances of circuits. In the second section, we shall consider the ability to localize the laser irradiation to small areas for realizing chemical surface reactions in the micron scale. The last section is devoted to the use of lasers for the recrystallisation of small-grain polycrystalline Si (Poly-Si) films into large area crystalline Si films for Silicon-on-Insulator (SOI) technologies and more importantly for future three-dimensional (3-D) circuitry.

I. LASER PROCESSING

Laser processing uses laser radiation to perform thermal processes involved in device fabrication. In the process flow, several steps require a high temperature treatment, these are (1) :

- Oxidation of Si for dielectric formation necessary in field effect devices (MOSFET's) as well as in isolation purposes.
- Annealing of implanted damage. Ion implantation is a well controlled method for doping Si. This step involves more or less damage to the Si lattice (depending on energy and dose used). Heating is necessary to recover the original structure and to activate the implanted ions by allowing them to migrate to substitutional sites.
- Diffusion of dopants. Another method for doping Si material is

Soares, O.D.D., Perez-Amor, M. (eds), Applied Laser Tooling. ISBN-13: 978-94-010-8096-5
© *1987. Martinus Nijhoff Publishers, Dordrecht.*

to allow dopant atoms from a source (spin-on glass) to migrate into Si. This diffusion (migration) is thermally activated. In the previous case with ion implantation, this diffusion leads to a spreading of the initial doping profile if the annealing is performed in a conventional furnace as shown in Fig. 1.

- Silicide formation. Si can react with transition metals (for ex. tungsten or Cobalt) forming silicides. The compounds are of interest for interconnects. Silicide formation is activated by thermal energy.
- Reflow of deposited oxide. Before the metallization step for interconnects, devices are covered with a deposited oxide. This oxide must be reflown at high temperature to taper edges so that interconnection lines can cross smoothly without being disrupted.

All the mentioned processes involve heating of the entire wafer in a furnace for long periods of time. The high temperature and the long exposure time favor i) The introduction of contaminants from the ambient into Si, ii) the dissolution and activation of metallic impurities previously in inactive phases.

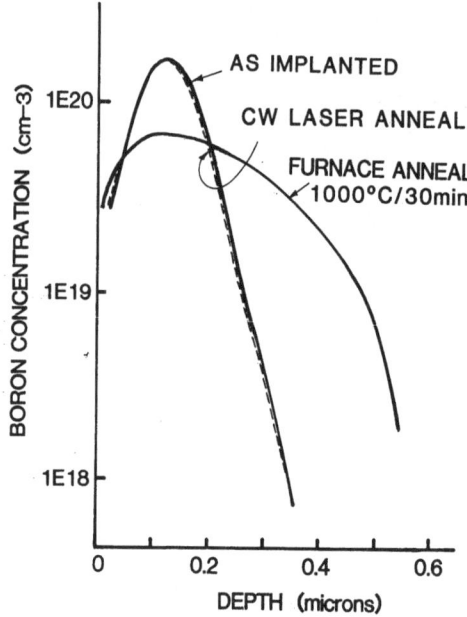

Fig. 1 : Distribution of the boron concentration with depth position obtained by SIMS profiling for as implanted, furnace anneal and Argon laser anneal. Implantation conditions : 2×10^{15} cm^{-2}, 35 Kev.

In current production lines, contamination in furnaces is under tight control. The most critical problem for several applications is the unwanted diffusion of dopant. In the case of an enhancement mode MOS transistor shown in Fig. 2, the diffusion results i) in parasitic capacitances reducing the operating speed ii) in a reduction of the electrical channel length leading in turn to an increase of leakage current, a decrease of punch-through voltage and a decrease of threshold voltage.

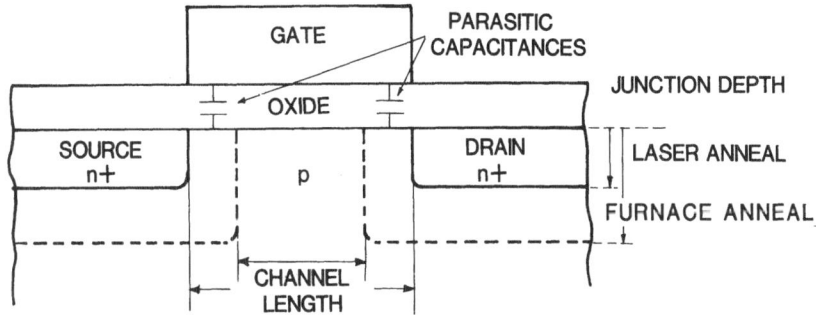

Fig. 2 : Effects of dopant diffusion on MOSFET operation.

Lasers provide a solution to these problems because :

. Heating takes place in delimited small areas (the focused spot diameter).
. Heating can be confined in depth in a small surface layer ;
. The heating cycle (heating and cooling down) is accomplished in a very short time (ms).

The fact that heating is localized in plane is the basis of laser microchemical processes discussed in section II.

Heating can be controlled in depth to the active surface layer of interest by the choice of the laser wavelength (the absorption depth depends on the wavelength) and the dwell-time for cw laser or the pulse-width for pulsed lasers. This property is used in the recrystallization of poly-Si in the fabrication of stacked circuits (cf. section III).

The short heating time limits the possibility of contamination and reduces the change in doping profiles (see Fig. 1). Note that laser treatment may also induce defects. Deep levels due to metallic impurities are found in the band gap after laser irradiation (2). The short heating cycle can produce unique non-equilibrium structures. By ion implantation one can introduce a large amount of dopant atoms in a thin surface layer. With pulsed laser annealing, the very fast cooling rate can freeze-in the profile allowing an activated impurity concentration exceeding the solid solubility limit of the dopant in Silicon. This property may be used to make good ohmic contacts and abrupt junction diodes (3).

To increase operating speed and packing density, the trend is to design and fabricate devices with smaller and smaller feature sizes. Channel length is now approaching the submicron range. Such submicron channel transistors cannot be realized with conventional furnaces. One or two micron channel length is the limit of furnace processing due to the previously mentioned diffusion of dopant.

With almost no diffusion of dopant obtained with laser processing one can expect better results. Indeed, a study of ring oscillators made in Si-on-sapphire shows that propagation delay time is faster by 25 % for the laser case compared to furnace (4). This results from reduced parasitic capacitances and an increased mobility.

Laser treatment also shows better results than furnace annealing for some specific structures such as :

. Activation of dopant in poly-Si for interconnects (5-7).
. Reduction of surface asperities of poly-Si before oxidation to reduce local field enhancement for MOS capacitors on poly-Si (8).
. Reduction of contact resistance and interdiffusion of contact constituants (5).
. Reflow of deposited glass (PSG) (9).
. Formation of silicides (10).
. Gettering effect by back-side laser induced damage (11).

Although lasers show a wide range of applications in device fabrication, they are not yet actually used in production lines. The reasons are :

. The low throughput due to the small spot size (the laser beam needs to be focused).
. The poor reproducibility (mostly with pulsed lasers).
. And the advent of graphite heater (12) and lamp systems (13).

Practically, the graphite or lamp systems perform heat treatment of the entire wafer within some tens seconds with results similar to those obtained with lasers concerning for exemple the redistribution of dopants, the reflow of PSG, the formation of silicide (12), etc.

II. LASER MICROCHEMICAL REACTIONS

The local character of laser treatment is used in a new technique : laser direct-write processing. No mask is needed.

Figure 3 shows schematically the experimental set-up for direct laser writing.

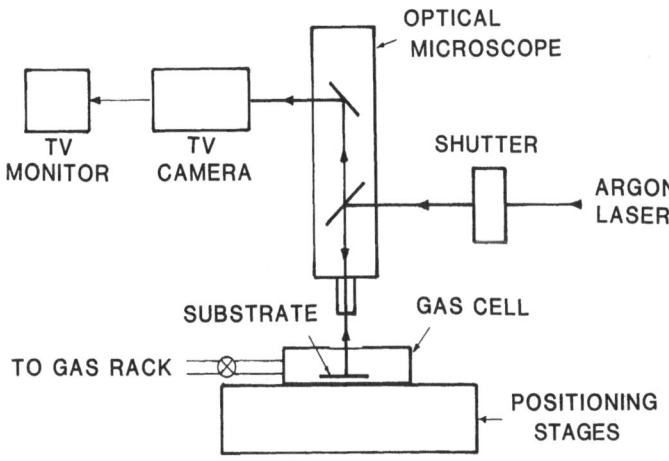

<u>Fig. 3</u> : Set-up for direct laser writing.

A laser beam can drive a chemical reaction on a surface (14). Basically, with a focused visible or u.v. laser beam, one can initiate :

- A photochemical reaction in the vapor adjacent to a surface. Photons can break molecular bonds initiating reactions at ambient temperature.
- A thermal reaction in an adsorbed molecular layer on the surface, the laser heats the substrate. This is classical chemical vapor deposition (CVD) but in a micron scale.

This last type of reaction is the most useful for technology. As reactions occur within a tightly focused beam of diamet er down to 2 microns, an extremely high resolution is obtained (better than 0,5 micron (15)).Due to a large variety of available reactions, a wide range of surface modifications is possible. One can etch, dope or deposit materials at the surface of Si. With these three techniques in hand, devices can be fabricated by using solely direct laser writing. MOS transistors have been actually made (16).

The laser microchemical techniques are particularly useful for precise surface modifications without the use of photomasks and multistep processing. Small modifications such as addition or deletion of structures can be made rapidly in one step. These techniques are helpful for fabrication fault correction. Increase in integration density results in reduction in yield due to random defects introduced during fabrication. A frequent type of fault is metallization shorts : this fault may be repaired by laser techniques (15, 17). Another important application of direct laser write technique is the restructuring of circuits for testing and evaluation of complex prototypes (18). High density devices such as a 64 K static RAM are very susceptible to yield loss due to failed elements. Redundant elements are then designed into these circuits. So the laser can disconnect failed circuit elements by cutting appropriate links and substitute spares for faulty elements (19).

The next application of laser direct write is thin-film resistor trimming by using a pulsed laser directly on wafers. The resistor is adjusted until desired results are obtained. This trimming technique is actually used in production line.

A new technique under investigation is the combination of a computer aided design system with a pulsed Yag or cw Argon laser for fabricating photomasks (20).

Concerning photolithography, several experiments have shown that excimer lasers (emissions in far u.v.) can advantageously replace classical mercury lamps. Resolution is better (shorter wavelengths) and exposure time shorter (high power available) (21).

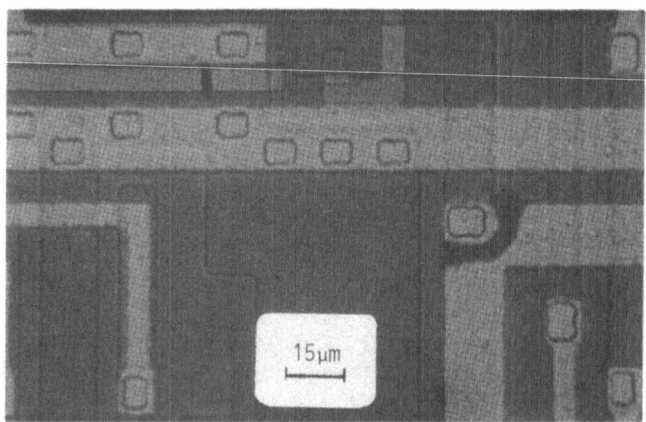

Fig. 4 : Example of circuit repairing. A contact pad is isolated without disrupting the Aluminium line (17).

Recent experiments show that excimer lasers can simultaneously expose and develop some photoresists (22, 23) and silicon monoxide (24). The molecular chains of the photoresist material are broken into small fragments by the u.v. photons. These fragments are then ejected out of the surface by absorbing the excess photons (24).

III. LASER RECRYSTALLIZATION OF POLYCRYSTALLINE SILICON FOR SILICON-ON-INSULATOR TECHNOLOGIES AND THREE-DIMENSIONAL CIRCUITS

A SOI structure consists of a thin Si layer on a insulating substrate. The latter can be a quartz plate, a sapphire plate or an oxide grown on a Si wafer.

SOI circuits are widely studied due to their dramatic advantages over bulk counterparts in packing density, in reduction of parasitic capacitances, in complete suppression of latch-up problem for CMOS circuits and in radiation tolerance.

Laser recrystallization of poly-Si deposited on oxide coated Si wafers is now one of the most studied technique for fabricating SOI structures. Figure 5 is a schematic of this structure with typical thicknesses encountered.

If after fabricating devices in the Si substrate by a standard technology, one deposits an oxide layer and on top of it a device-worthly Si layer is realized in which devices are again made (without damaging the previous ones) a 3-D structure is obtained.

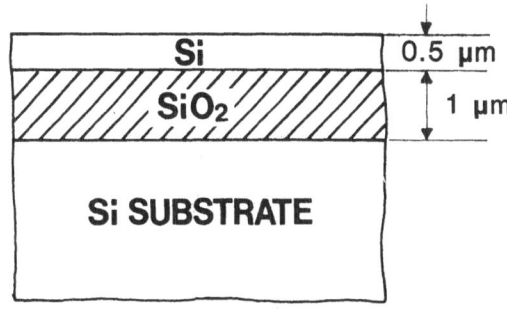

Fig. 5 : Schematic cross-section of the SOI structure made on a Si wafer.

Laser recrystallization techniques

Pulsed lasers, the first lasers used for Poly-Si recrystallization, can only bring about an increase in grain size of up to a few microns. This kind of laser is no longer used in film growth experiments. Cw CO_2 lasers are used for recrystallization of silicon films on quartz substrates (25), but their long wavelength precludes superficial heating required for 3-D circuit fabrication. The most widely used laser for SOI and 3-D processing is the cw argon laser. Owing to its 500 nm wavelength, energy is deposited only within a thin superficial layer. Hence, melting and recrystallization of a polysilicon film can be obtained without raising temperature in the underlying active layers above 900°C if a 1 μm-thick oxide layer is intercalated between silicon layers. Furthermore, the short dwell time (< 1 ms) involved implies virtually no dopant diffusion within the underlying devices. Within the last four years, many improvements have been made in laser recrystallization, all of which aim at producing defect-free silicon films or islands.

The key for growing large (single-) crystals during laser recrystallization is controlling the shape of the crystallization interface (trailing edge of the molten zone).

The beam profile of a cw laser being gaussian, chevron-like recrystallization patterns are observed after scanning a Poly-Si sample with a conventional focused circular laser beam (Fig. 6a). This pattern is caused by the convex shape of the trailing edge in the wake of the advancing spot. Crystal growth proceeds along the thermal gradient (perpendicular to the trailing edge), i.e. crystallites grow from the edges of the scanned line towards its center. When a concave trailing edge is produced, crystal growth proceeds outwards from the alrea dy grown crystal towards the edges of the scanned line. Random nucleation inwards from the edges is thus ruled out (Fig. 6b).

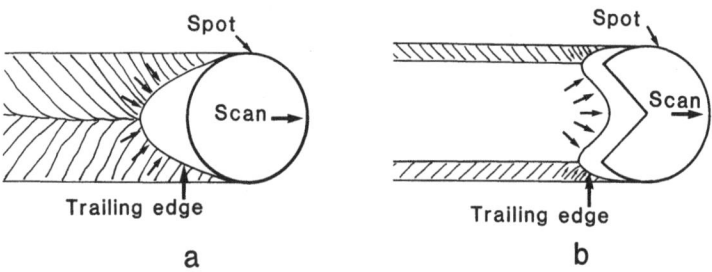

Fig. 6 : Chevron structure obtained with a gaussian laser spot (a) and large grain growth using a partly masked beam (b).

A concave trailing edge can be obtained by shaping either the laser beam (26-28) or the thermal response of the sample through oxide layer thickness variations (29-31), a dummy polysilicon layer (32), an antireflection coating (33, 34). This last method which provides excellent and reproducible results, is the patterning of an antireflection cap on top of the silicon film to be recrystallized. Figure 7 shows how tailoring of the trailing edge is achieved using stripes of an antireflecting layer (SiO_2, Si_3N_4, or Si_3N_4 on SiO_2). In this case, the microfloating zone is composed of a succession of concave interfaces. All defects, among them grain-boundaries, are then swept underneath the antireflecting stripes by the thermal gradient, and defect-free crystals are obtained, with localized defects between them. It is interesting to note that the patterning is made by photolithographic techniques, thus the location of the defects within the chip is known and devices can be placed in defect-free areas. This technique has already been used to recrystallize active layers of 3-D integrated circuits (35).

Further improvements on the method have been obtained by using a raster scan of the laser perpendicular to the antireflection stripes in order to induce a pseudo linear advancing molten zone (36-38). By opening a seeding window at the beginning of the SOI structure, <100> crystal areas 100 μm long and 5 mm wide can be grown in which the location of the remaining defects (subboundaries) is fully controlled (Figs. 8-9). These defects can be removed by etching or selective oxidation during further processing steps (37). Patterning of the antireflection coating is not limited to stripes. Other antireflection patterns (grids, squares) have recently been demonstrated to provide single-crystal areas (38).

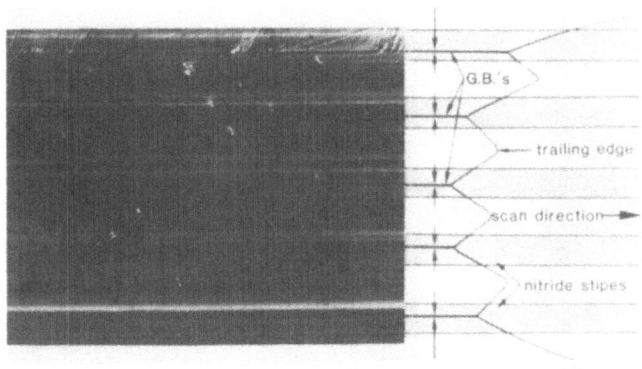

Fig. 7 : Tailoring of the trailing edge and large crystal growth (left to right scan of the laser beam). The spacing between grain boundaries is 20 microns.

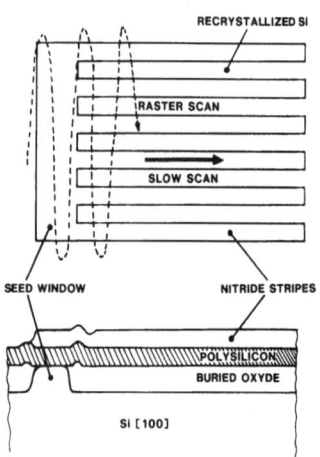

Fig. 8 : Oscillatory scan and seeding window used
to grow large single-crystals.

Fig. 9 : Etch pit grid pattern of the recrystallized film using
the raster scan technique showing that the recrystallized
Si layer has a single crystalline orientation.
The film is also decorated with Secco etch to reveal grain
and subgrain boundaries. The distance between etch pits is 30 μm.

The most dramatic defects found in laser-recrystallized silicon films are grain boundaries. When located in the transistor channel area, a grain boundary parallel to the current flow can induce a short-circuit between source and drain because of enhanced dopant diffusion along the grain boundary (40). When the boundary is perpendicular to the carrier flow, it brings about an uncontrolled shift of the threshold voltage and a decrease of the mobility of the carriers in the channel (41). Subgrain boundaries, which are simply dislocation networks between slightly misoriented crystals (42), seem to have a much weaker electrical activity (43).

Devices have been made to determine the electrical properties of this laser recrystallized Poly-Si using the antireflection stripes and the raster scan techniques. Figure 10 shows typical $I_D(V_G)$ characteristics of a n-channel transistor made in a defect-free area. The results of such measurements show that the recrystallized Si has electrical characteristics approaching those of bulk Si.

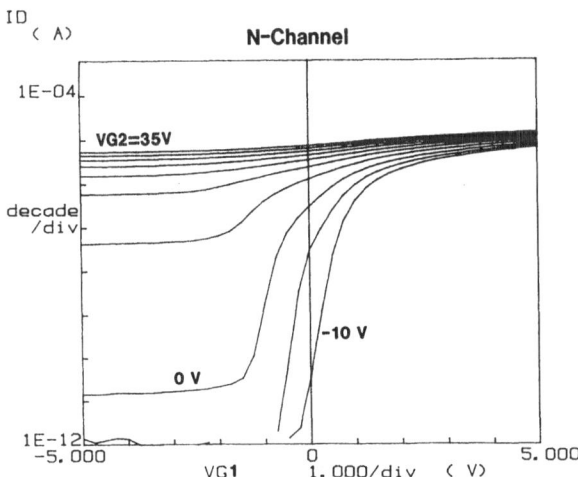

Fig. 10 : Experimental $I_D(V_{G1}, V_{G2})$ curves obtained with a n-channel device made in defect-free laser recrystallized Si. L = 6 μm, Z = 20 μm. G_1 : upper gate, G_2 : lower gate, upper gate oxide : 100 nm, lower gate oxide : 1 μm, $N_a = 5 E 15 cm^{-3}$.

3-D integration : State of the art

Many different approaches to 3-D integrated structures have been reported in the literature, all based on two key issues : growth of device-worthy silicon films on an amorphous insulator (typically SiO_2) and recrystallization of devices in upper layers that does not damage the underlying structures already fabricated.

The different results published up to now can be classified in the following way :

a) Stacked CMOS, in which a single gate is shared by the n and the p-channel transistor (44-46). The channels of both the upper and the lower transistor are controlled by a single gate (Fig. 11). This merging capability of stacked CMOS structures offers a large increase in packing density with respect to conventional bulk CMOS (48). 64 K CMOS SRAMs have been made using this stacking principle (49). In the stacked transistor CMOS approach (STCMOS), all n-channel transistors are placed in the substrate and the p-channel devices are made in the recrystallized layer, which is of poorer quality than the substrate. STCMOS devices are therefore best adapted for nMOS-oriented CMOS technologies.

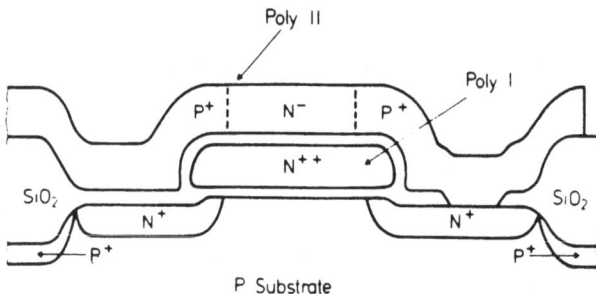

Fig. 11 : Cross section of a stacked CMOS inverter (47).

b) Multilayer SOI, in which SOI layers are stacked upon one another, the 0-th layer being the substrate. The different layers are interconnected using a classical metallization process and there is no field-effect interaction between devices belonging to different layers. Thick and planarized isolation dielectrics between active layers allow growth of high quality single-crystal by the laser technique without any degradation of the underlying devices (Fig. 12) (35, 50-52).

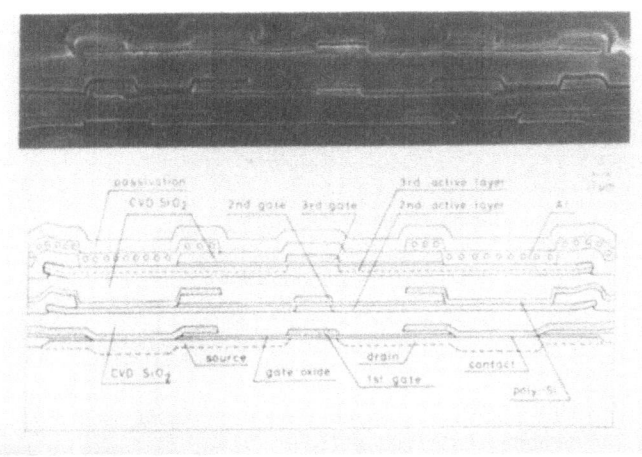

Fig. 12 : SEM and schematic cross-section of a SOI/SOI/bulk
silicon triple layer structure (35).

c) Optical sensor devices, such as a photodetector array made in recrystallized silicon and its integrated optical waveguide (53) or the 10-bit linear image sensor fabricated in double active layers with the photodiodes in the top laser-recrystallized SOI layer and the signal processing circuits in the bottom bulk silicon (54).

d) Staggered CMOS (55), which is devoted to the realization of specific devices such as SRAM cells. This approach provides very dense structures in which the drain of one transistor serves as gate for another, and so on. The process is unfortunately incompatible with standard technologies. Indeed, the staggered CMOS process involves a high energy, high dose implantation step to create the bulk source and drain regions.

e) Double gate devices, like the Cross-MOS inverter (56) where a single active layer is controlled by two gates, one over it and the other below it. The Cross-MOS inverter offers poor transfer characteristics and its process is not standard.

256

f) "Mezzanine" devices, in which part of the structure is made in the bulk and part of it is made in the laser recrystallized SOI. Here advantage is taken from the 3-D arrangement of different silicon layers to create dense, high-performance DRAM cells (Fig. 13) (57-60).

LASER RECRYSTALLIZED SI

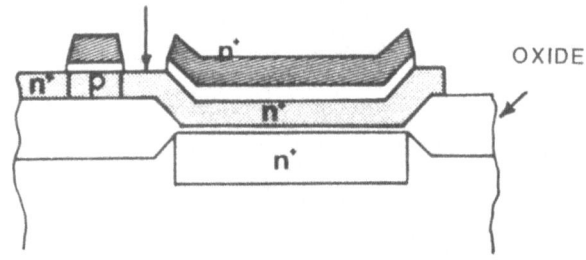

Fig. 13 : Dynamic RAM cell made in laser recrystallized silicon (57).

It is worth noting that although there are many techniques for producing device-worthy SOI films (ELO, SIMOX, FIPOS, various beam recrystallization techniques), with the exception of passivated polysilicon devices (49), all the existing 3-D devices have been realized using a laser recrystallization step. Indeed, laser recrystallization is the only technique providing device-worthy silicon films on an already processed substrate that neither induces overheating nor damages devices fabricated in the underlying layers, owing to the unique features of superficial heating and short time involved. E-beam recrystallization is quite similar to laser annealing, but it induces damage in the underlying layers (61). Low temperature solid phase lateral epitaxial growth of silicon looks very promising for 3-D IC fabrication, but this technique is only in the early stages of development and should be considered for long term projects only.

SUMMARY

Laser radiations can perform a variety of thermal processing in semiconductor technology.

The laser techniques offer real advantages for fabrication of circuits having submicron feature sizes. Shortcomings with lasers are the low throughput, and the poor reproducibility. A new application of lasers is opened up with micro-"surgery" of integrated circuits. This is particularly useful for development and testing of high density ICs.

The most important application of lasers in IC technology is the conversion of small grain poly-Si to single-crystal without affecting the characteristics of underlying layers. This leads to SOI and 3-D structures.

ACKNOWLEDGMENTS

I would like to thank J.P. Colinge, J.C. Pfister, V.T. N'Guyen for helpful discussions, A. Gonthier for typing the manuscript.

REFERENCES

(1)   Sze S.M., VLSI Technology, Mc GrawHill Intern. (1983).
(2)   Chantre A., Appl. Phys. Lett. 46, 263 (1985).
(3)   Stuck R., E. Fogarassy, J.C. Muller, M. Hodeau, A. Wattiaux and P. Siffert, Appl. Phys. Lett. 38, 715 (1981).
(4)   Yaron G. and L.D. Hess, IEEE Trans. Electron. Devices ED-27, 573 (1980).
(5)   Hess L.D., G. Eckhardt, S. A. Kokorowski, G.L. Olson, A. Gupta, Y.M. Chi, J.B. Valdez, C.R. Ito, E.M. Nakaji and L.F. Lou, Laser-Solid Interactions and Transient Thermal Processing of Materials, Eds J. Narayan, W.L. brown, R.A. Lemons, North-Holland, 337 (1983).
(6)   Ternisien d'Ouville T., D.P. Vu, A. Perio and A. Baudrant, J. Appl. Phys. 53, 5086 (1982).
(7)   Calder I.D. and H.M. Naguib, IEEE Electron. Dev. Lett. EDL-6, 557 (1985).
(8)   Yaron G., L.D. Hess and S.A. Kokorowski, IEEE Trans. Electron Devices ED-27, 964 (1980).
(9)   Achard H., E. Regent, J.P. Joly, J.M. Hode and P. Jeuch, Electrochemical Society Meeting, Oct. 9-14 (1983), Abstract 231, 365.
(10)   Bomchil G., D. Bensahel, A. Golanski, F. Ferrieu, G. Auvert, A. Perio and J.C. Pfister, Appl. Phys. Lett. 41, 46 (1982).
(11)   Sandow P.M., Solid State Technology, July 80, p. 74.
(12)   Fulks R.T., C.J. Russo, P.R. Hanley and T.I. Kamins, Appl. Phys. Lett. 39, 604 (1981).
(13)   Gat A., IEEE Electron. Dev. Lett. EDL-2, 85 (1981).
(14)   The reader can find the details of the chemical reactions induced by a laser beam in the chapter by D. Bäuerle. We mention in this paper only applications of laser-chemical reactions in microelectronics.
(15)   Tsao J.Y. and D.J. Ehrlich, Appl. Phys. Lett. 43, 146 (1983).
(16)   McWilliams B.M., I.P. Herman and F. Mitlitsky, R.A. Hyde

258

        and L.L. Wood, Appl. Phys. Lett. <u>43</u>, 946 (1983).

(17)   Auvert G., private communication.

(18)   Ehrlich D.J., J.Y. Tsao, D.J. Silversmith, J.H.C. Sedlacek, R.W. Mountain and W.S. Graber, IEEE Electron. Dev. Lett. <u>EDL-5</u>, 32 (1984).

(19)   Smith R.T., J.D. Chlipala, J.F.M. Bindels, R.G. Nelson, F.H. Fisher and T.F. Mantz, IEEE J. Solid State Circuits <u>SC-16</u>, 506 (1981).

(20)   Swenson E.J., Solid State Technology, Nov. 1983, p. 156

(21)   Jain K., C.G. Wilson and B.J. Lin, IEEE Electron. Dev. Lett. <u>EDL-3</u>, 53 (1982).

(22)   Srinivasan R. and V. Mayne-Banton, Appl. Phys. Lett. <u>41</u>, 576 (1982).

(23)   Geis M.W., J.N. Randall, T.F. Deutsch, P.D. DeGraff, K.E. Krohn and L.A. Stern, Appl. Phys. Lett. <u>43</u>, 74 (1983).

(24)   Fiori C. and A.B. Devine, Appl. Phys. Lett. <u>47</u>, 361 (1985).

(25)   Weinberg Z.A., Appl. Phys. Lett. <u>39</u>, 421 (1981).

(26)   Kawamura S., J. Sakurai, M. Nakano and M. Takagi, Appl. Phys. Lett. <u>40</u>, 394 (1982).

(27)   Stultz T.J. and J.F. Gibbons, Appl. Phys. Lett. <u>39</u>, 498 (1981).

(28)   Hode J.M., J.P. Joly and P. Jeuch, Techn. Dig. of ECS Spring Meeting, Montreal (1982) 232.

(29)   Biegelsen D.K., N.M. Johnson, D.J. Bartelink and M.D. Moyer, Appl. Phys. Lett. <u>38</u>, 150 (1981).

(30)   Possin G.E., H.G. Parks, S.W. Chiang and Y.S. Liu, Proc. of IEDM, 424 (1982).

(31)   Kawamura S., S. Sasaki, M. Nakano and M. Takagi, J. Appl. Phys. <u>55</u>, 1607 (1984).

(32)   Mukai R., S. Sasaki, T. Iwai, S. Kawamura and M. Nakano, Proceedings of IEDM, 360 (1983).

(33)   Colinge J.P., E. Demoulin, D. Bensahel and G. Auvert, Appl. Phys. Lett. <u>41</u>, 346 (1982).

(34)   Colinge J.P., E. Demoulin, D. Bensahel and G. Auvert, Jpn. J. Appl. Phys. Suppl. <u>22-1</u>, 205 (1982).

(35)   Nishimura T., K. Sugahara and Y. Akasaka, FED SOI/3D Workshop, March 19-21 (1985) Shuzenji, Japan.

(36)   Colinge J.P., D. Bensahel, M. Alamome, M. Haond and J.C. Pfister, Electronics Letters <u>19</u>, 985 (1983).

(37)   Colinge J.P., D. Bensahel, M. Alamome, M. Haond and C. Leguet, Energy-beam Interactions and Transient Thermal Processing, Ed. by J.C.C. Fan and N.M. Johnson, North-Holland, New-York, 597 (1984).

(38)   Drowley C.I., P. Zorabedian and T.I. Kamins, Energy-beam Interactions and Transient Thermal Processing, Ed. by J.C.C. Fan and N.M. Johnson, North-Holland, New-York, 465 (1984).

(39)   Iwai T., S. Kawamura and M. Nakano, Abstracts of the Fall Meeting of the MRS 1983, paper A7.19, 57 (1984).

(40)   Ng K.K., G.K. Celler, E.I. Povilonis, R.C. Frye, H.J. Leamy and S.M. Sze, IEEE Electron. Dev. Lett. 2, 316 (1981).

(41)   Colinge J.P., H. Moel and J.P. Chante, IEEE Trans. on Electron. Devices 30, 197 (1983).

(42)   Haond M., D.P. Vu, D. Bensahel and M. Dupuy, J. Appl. Phys. 54, 3899 (1983).

(43)   Vu D.P., A. Chantre, H. Mingam and G. Vincent, J. Appl. Phys. 56, 1682 (1984).

(44)   Gibbons J.F. and K.F. Lee, IEEE Electron. Dev. Lett. 1, 117 (1980).

(45)   Goeloe G.T., E.W. Maby, D.J. Silversmith, R.W. Mountain and D.A. Antoniadis, Proceedings of IEDM, 544 (1981).

(46)   Robinson A.L., D.A. Antoniadis and E.W. Maby, Proceedings of IEDM, 530 (1983).

(47)   Colinge J.P., E. Demoulin and M. Lobet, IEEE Trans. on Electron. Dev. 29, 585 (1982).

(48)   Hoefflinger B. S.T. Liu and B. Vajdic, IEEE J. of Solid-State Circuits SC-19, 37 (1984).

(49)   Malhi S.D.S., R. Karnaugh, A.H. Shah, L. Hite, P.K. Chatterjee, H.E. Davis, S.S. Mahant-Shetti, C.D. Gosmeyer, R.S. Sundaresan, C.E. Chen, H.W. Lam, R.A. Haken, R.F. Pinizzotto and R.K. Hester, Presented at the Device Research Conference, Santa Barbara (1984) paper VB-1.

(50)   Kawamura S., N. Sasaki, T. Iwai, R. Mukai, M. Nakano and M. Takagi, Proceedings of IEDM, 364 (1983).

(51)   Akiyama S., S. Ogawa, M. Yoneda, H. Yoshii and Y. Terui, Proceedings of IEDM, 352 (1983).

(52)   Nishimura T., K. Sugahara, Y. Akasaka and H. Nakata, Ext. Abstr. of the Conf. on Solid State Devices and Materials, Kobe, 527 (1984).

(53)   Wu R.W., H.A. Timlin, H.E. Jackson and J.T. Boyd, Appl. Phys. Lett. 46, 498 (1985).

(54)   Hirose S., T. Nishimura, K. Sugahara, S. Kusunoki, Y. Akasaka and N. Tsubouchi, Symposium on VLSI Techno., May 14-16 (1985) Kobe, Japan.

(55)   Maby E.W. and D.A. Antoniadis, Paper presented at the MRS Spring Meeting, Albuquerque (1984).

(56)   Gibbons J.F., K.F. Lee, F.C. Wu and G.E.J. Eggermont, IEEE Electron Dev. Lett. EDL-3, 191 (1982).

(57)   Jolly R.D., T.I. Kamins and R.H. McCharles, IEEE Electron. Dev. Lett. EDL-4, 8 (1983).

(58)   Sturn J.C., M.D. Giles and J.F. Gibbons, IEEE Electron. Dev. Lett. EDL-5, 151 (1984).

(59)   Gibbons J.F. and K.F. Lee, Proceedings of IEDM, 111 (1982).

(60)   Ohkura M., K. Kusukawa, H. Sunami, T. Hayashida and T. Tokuyama, Proceedings of IEDM, 718 (1985).

(61)   Saitoh S., K. Higuchi and H. Okabayashi, Jpn. J. Appl. Phys. Suppl. 22-1, 197 (1983).

OPERATION OF A LASER FACILITY :

MAINTENANCE - IN PROCESS DIAGNOSTICS - INVESTMENT AND RUNNING COSTS

by Alberto SONA

Director of the Italian CNR Program on High Power Lasers

Center for Information,Studies and Experiments, (CISE)
via Reggio Emilia N 39 - 20090 SEGRATE (MILANO) ITALY

INTRODUCTION

The operation of a laser facility is much similar to the operation of other industrial equipments, the special procedures typical of laser being a limited part of the total.

In modern lasers the operation is fully automated and a laser workstation with loading, positioning and unloading under numerical control can be operated by a single supervising technician.

The laser needs supply of electricity, active gas mixture, assistance gas for the process, cooling water. Only gases are usually supplied discontinuosly in bottles requiring the assistance of the operator.

Maintenance refers to the parts to be revised and/or replaced at longer time intervals and appropriate schedules are to be followed according to the manufacturers instructions.

Checking and monitoring the laser beam along its path is one of the most important tasks of the laser operator and it is related both to the laser itself and to the external beam delivery system.

In the following a short outline of typical maintenance and checking problems is given followed by some examples on the running costs of a laser workstation.

MAINTENANCE AND MONITORING

The laser workstation has conventional electric and mechanical components such as power supplies, blowers, pumping units and specific components for the excitation of the active medium and laser power extraction. In gas lasers the active medium is a flowing gas mixture where an electrical discharge is maintained between a set of electrodes. Due to the ions bombardment the cathode surface slowly deteriorates and various processes of oxidation or coating with lower conductivity layers occurs requiring the cleaning or the substitution of the cathode. Copper cathodes usually require maintenance every 300 - 500 hours depending on the power level. The cleaning or substitution usually can be performed in about one hour. Maintenance cycles can therefore be arranged on a weekly basis even for a two shifts operation.

Soares, O.D.D., Perez-Amor, M. (eds), Applied Laser Tooling. ISBN-13: 978-94-010-8096-5
© *1987. Martinus Nijhoff Publishers, Dordrecht.*

Solid states laser do require flash-lamps substitution typically after 10 million shots which means, at the usual 10 pulses / second rate , about 300 working hours again allowing a maintenance cycle on a weekly basis.The replacement of a set of lamps can be usually accomplished in a time of the order of one hour.In both cases the continuous monitoring of electrical parameters allows an early prediction of the wear condition of these components.

Optical power extraction is accomplished by a set of mirrors forming the optical cavity. Maintenance requirements are twofold :

a - Mirrors align ment optimization which can be performed in real time by the operator; actually many lasers have remotely adjustable mirrors or even servo-controlled alignement systems. This requires rather unfrequent corrections in well designed lasers and it is a minor maintenance problem. Actually most laser units up to the 1 kW level are designed with stable and reliable structures not requiring at all realignement for time intervals of months.

b - Mirror replacement due to surface deterioration is sometimes required. The extremely high power densities inside the cavity can give rise, often in connection with the presence of circulating dust particles, moisture or scratches on the mirrors to surface damage. Spare mirrors must always be available; the substitution requiring again times of the order of two hours. A mirror lasts typically for more then 2000 hours, the life depending strongly on the cleanless of the enviroment. Mirrors can be cleaned but if the surface is severely damaged reworking is necessary. For copper mirrors diamond turning machines can restore the surfaces in times of the order of a few hours.Glass mirrors reworking similarly has to be made at the factory and spare parts are needed.Modern lasers have all the optical components in a protected environment and in most CO_2 lasers the mirrors surfaces are mounted inside the same vessel where the gas mixture is contained and only one output optical port is provided.

The maintenance of external optics is easier ;the most critical component being usually the focusing lens or mirror which is protected by the coverage gas injected on the workpiece and sometime by an additional gas jet. Occasionally debris or small molten metal droplets may hit the optical surfaces originating local scratches.Howeverin well designed and operated equipments the external optics are also long lived as they are exposed to a smaller power intensity. Life in excess of 1000 hours can be expected , the actual value depending again on the cleanness of the environment and on the effectiveness of the gas shield.

Modern lasers have built - in self checking facilities usually provided by a microprocessor.Flow diagrams and operational parameters can be displayed by a suitable software on the monitor screen providing information on the gas flow path, the electrical discharge parameters, the optical beam path and alignement state ,the modal structure, the external optics alignement state.

All these facilities are usually provided in sophisticated high power units intended for operation in optimized conditions with maximum efficiency.As a conclusion a laser workstation does not require more complex maintenance than usual tooling machines do

MANAGEMENT AND COSTS OF A LASER WORKSTATION

The hourly cost of a laser depends on the equipment cost, consumables costs and operator salary. In fig 1 the cost of CO_2 lasers versus power is reported resulting in an average figure of 80 $/watt for units in the range of a few kilowatts.A similar information is given in fig 2 for solid state lasers having a cost of 400 $/watt of average power. The factor five betweeen the two costs is mainly depending on the lower efficiency of the solid state lasers (0.02 instead of 0.10).

As regards consumable materials the typical operating cost of a 2.5 kW CO_2 laser for one year (2000 working hours) are the following :

a - Electrical energy	4.0	MLit.
b - Gases (includig laser mix and process)	5.0	MLit.
c - Cooling water	1.5	MLit.
d - Spare parts and components	5.5	MLit.
e - Maintenance - repair	2.0	MLit.
TOTAL	20.0	MLit.

The above figures scale almost linearly with power for points a,b,c,d.

In addition the operator salary has to be added amounting to about 30 MLit.

The capital cost of the laser and of the associated equipment can be estimated as follows :

Laser source (2.5 kW)	300 MLit.
Worktable and numerical control	100 MLit.
Installation (including water cooling facility, gas and electricity conn. etc.)	20 MLit.
TOTAL	420 MLit.

A well designed CO_2 laser will last for more than 8 years, however to evaluate the yearly cost one can use a three years period to cover the initial expenses . A further 5 years period is allowed for building up the capital for the purchase of the replacement. As a consequence the yearly costs are the following:

	2.5 kW	1.0 kW
Capital cost (420/3 - 150/3)	140 MLit.	50 MLit.
Running cost (2000 hours)	20 MLit.	10 Mlit.
Operator salary	30 MLit.	30 MLit.
TOTAL YEARLY	190 MLit.	90 MLit.
COST PER HOUR (2000 hours)	95 KLit.	45 KLit.

(1$ = 1700 Lit. ; 1 peseta = 12 Lit.)

For economic operation the laser workstation has to produce goods with an added value larger than the estimated cost. In practice factors of 2 up to 5 have been demonstrated allowing the return of the investment in less than one year. Additional advantages could be achieved running the workstation on a two shifts basis. The quoted values are of course only orientative but they are consistent with the data reported in ref.(1) where additional benefits where quoted due to the special taxation regime applied to investments in new high technology equipments.

Similar estimates can be performed for solid states lasers as reported f.i. in ref. (2).

In conclusion the laser hourly cost is mainly depending on the capital cost, the laser being a capital intensive technology. The operator cost and the running cost have a decreasing importance. The key point for a successfull economic operation is an effective exploitation of the laser workstation on a full shift or possibly on two shifts. The expected reductions in laser costs will enhance the diffusion of lasers in manufacturing which is at the present time mainly oriented towards high added value items. The cost estimate has in practice to be compared with the benefits of the laser processing which can be evaluated only with an accurate analysis of the advantages resulting from direct or indirect productivity and quality improvements.

REFERENCES

1) - B.G.Green et al. Proc. of LIM - 2 Birmingham 85 Publ.by IFS
2) - T.H.Weedon Proc. ICALEO 83 Los Angeles, pg. 24 Publ.by LIA

265

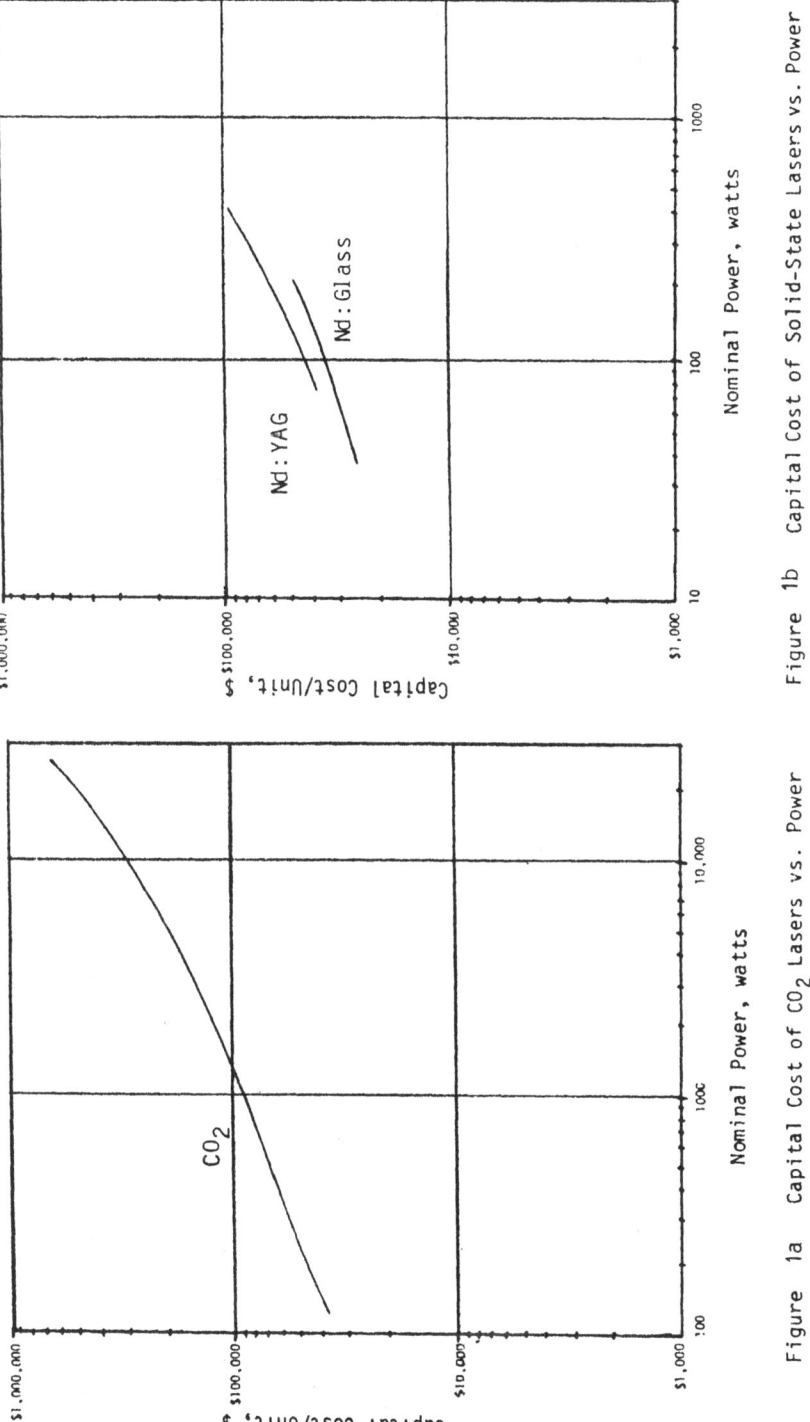

Figure 1b   Capital Cost of Solid-State Lasers vs. Power

Figure 1a   Capital Cost of CO_2 Lasers vs. Power

266

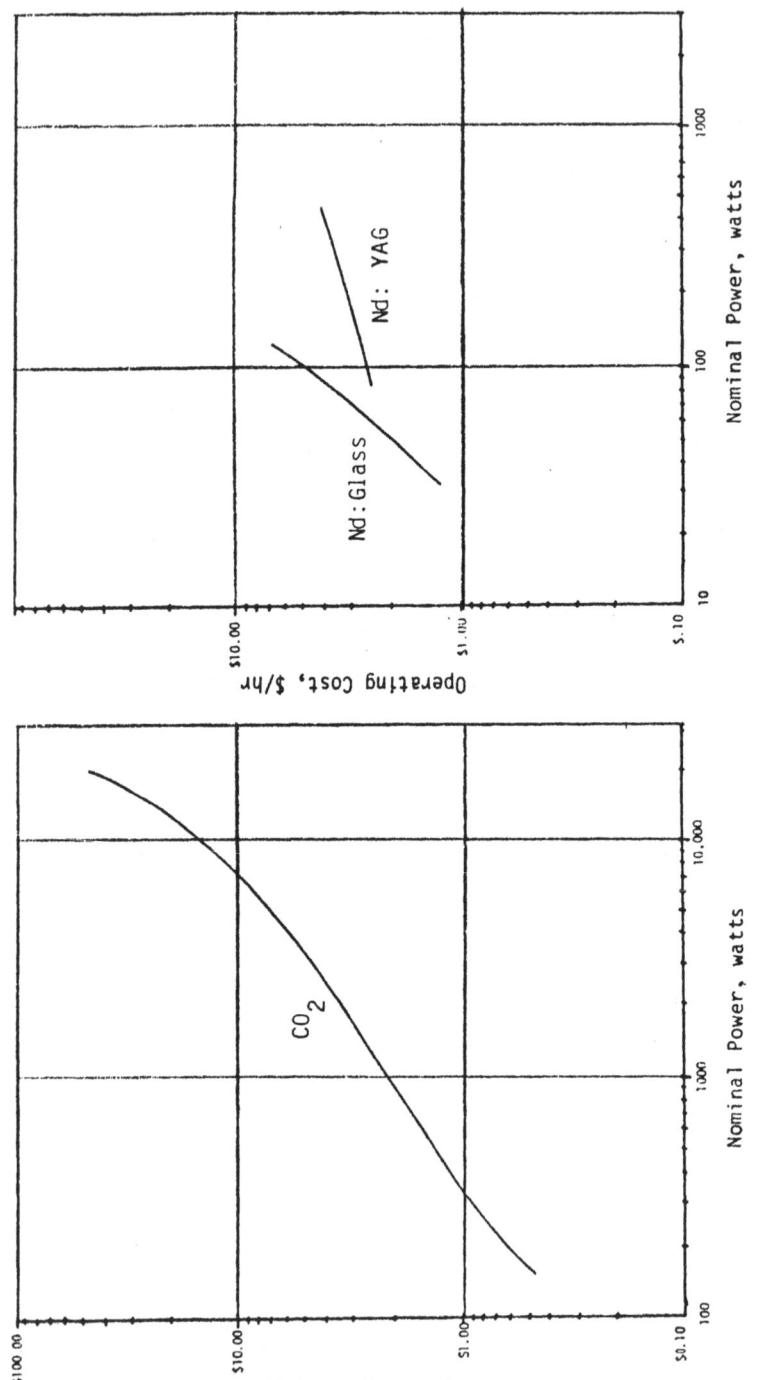

Figure 2-b Solid-State Operating Cost vs. Power

Figure 2-a CO₂ Operating Cost vs. Power

CURRENT AND FUTURE TRENDS IN LASER PROCESSING.

by W.M.Steen
Metallurgy and Material Science Department,
Imperial College of Science and Technology,
London SW7

Introduction

Currently the laser market in material processing systems is worth more than 350M$/year and has risen around 25% since last year.   The future is always difficult to predict but there are features which indicate the way the trend is set. Listing these indicators would include:
* Extrapolation of the market trend, which gives a 25% increase for a few more years.
* Extrapolation of the growth of new applications, which shows a similarly bouyant aspect.
* The fact that the laser has limited competition in the areas of high quality welds, cuts and surface effects.
* The fact that the laser is uniquely qualified as the energy source for automatic machinery.
* The fact that there is a firm belief amongst designers, financiers and scientists that the laser has a large part to play in manufacturing industry.
* The fact that government sponsored investment programmes around the world in laser technology are impressive. This must lead to significant developments in this technology.

These add up to one thing and that is that the laser has an expanding future for as far as one can reasonably predict.

In this chapter there is a consideration of the major growth areas and potential future growth areas as seen by the author.   Naturally any prediction is likely to err but may be this analysis, even if proved wrong, will stimulate some thoughts somewhere amongst the readers- if any!

The growth areas present and future are and will be associated with either the equipment or how it is used.   By analogy with computers this can be seen as the hardware and software of laser material processing.

Considering the equipment there are three major components:
* The laser
* The beam optics
* The process control systems

Considering the process and application developments there are again three aspects:

Soares, O.D.D., Perez-Amor, M. (eds), Applied Laser Tooling. ISBN-13: 978-94-010-8096-5

*   The type of application
*   Process improvements through better engineering and understanding
*   Fundamental improvements in understanding of the process mechanisms and structures produced.

These will now be considered in turn:

Developments in laser design:

Most processes work best at certain powers but on the whole there are usually advantages in having more power. Since with increased power productivity is likely to increase, heat affected zones will diminish with increased processing speed and a greater processing flexibility will also be gained. Thus there is a strong trend towards more powerful lasers. The current EUREKA programme in Europe has some targets which might include a 100kW CO_2 laser, a 10kW YAG laser and a 5kW Excimer laser. A number of manufacturers are now offering 10kW lasers (United Technology, Ferranti, Mitsubishi, Combustion Engineering). Development of 20kW lasers is in hand. Increasing power brings increased capital costs and usually a more temperamental machine. Thus the high powered argument may be solved another way. The Italians (1) have developed a beam guidance system for threading beams together. So a high powered beam can be built up by adding the beams from several small reliable modules. Imperial College has coupled two lasers together as an oscillator/amplifier pair (2). There is a floor space penalty to be paid for these approaches.
Some companies have solved this latter problem by offering a machine built on a mezzanine floor above the work station (3). Such an arrangement may also have the advantage of requiring one or two fewer guidance mirrors.
The power versus space problem is near the heart of recent successful designs and design thinking. Smaller power supplies and more compact cavity design for CO_2 lasers has been achieved by several companies using RF excitation instead of DC. (DFVLR,Trumpf, Laser Innovation Gmbh, Laser Corporation of America, Combustion Engineering MLI)(4,5,6). Some of these machines also address the problem of a more uniform mode structure by having the transverse excitation cavity on both sides of the fan in order to avoid the problem that is common to most transverse flow lasers that of an assymmetric power distribution due to the temperature gradient in the gas stream.
There are serious studies being made into the possibility of designing lightweight, high powered lasers with a view to being able to mount the laser head on an articulated arm robot (7).
Perhaps more fundamental in the design of lasers for material processing is the trend to investigate shorter wavelengths. The advantages of shorter wavelengths is that they are more readily absorbed by high reflectivity surfaces, such as copper, aluminium, gold and silver. Shorter

wavelengths can also, in theory, be more tightly focussed allowing high power densities from low powered machines. The spot size of a diffraction limited beam is proportional to the wavelength and the absorptivity is roughly described by Bramson's equation (8) to be inversely proportional to the wavelength. So there is good theoretical justification for investigating the possibilities of working with shorter wavelengths. Current research is tending towards 1kW YAG and 2-100W ultratviolet Excimer lasers. Attractive as these machines may appear there are some problems not currently discussed which may become dominant. These include the relatively safe nature of CO_2 10.6 μm radiation. This wavelength will not penetrate the eye and so energy densities some 10 times more intense can be tolerated with CO_2 radiation than can be with YAG or ultra violet since these are concentrated by the eye's own optics. Thus specular reflection from high powered ultraviolet lasers would have to be guarded against by suitable engineering.

Optics Design

One of the principal advantages of the laser in material processing is the ease with which optical power can be focussed and directed. There are, however, some particular difficulties with beam guidance in that it must not disturb the beam mode structure,power or positional accuracy. These design requirements can be summarised as follows:

* Beam mode structure must be preserved or improved by the guidance optics.

* Power losses should be negligeable.

* System must be strong enough to withstand thermal shock and point absorption defects caused by dust etc..

* The pointing accuracy must be better than + 25 μm

* There must be no vibration during movement.

* Guidance must be swift and controlled by fast software.

* The emergent beam must be axisymmetric for use in robotic applications. This means a circular beam cross section with no astigmatism from the guidance train. The beam must also be circularly polarised.

These conditions are difficult to achieve. However they do illustrate the areas of current development and those which are likely in the foreseeable future. Considering them in turn:

Control of mode structure:

Most material processing with a laser is best done with certain particular power distributions (9). Thus it is important that the mode is either preserved or improved by the guidance system. Various techniques for beam shaping have been known for some time such as integrating mirrors which shape the beam into squares of uniform power distribution. New optics are being investigated which are able to convert a poor mode into a near Gaussian mode. Such optics are specific to a particular laser and require considerable care in computing and manufacture. Variable optics which change their shape under pressure from piezo electric actuators are also under investigation (10).

Fibre optics, which are very attractive for beam guidance and the insertion of the beam into a manufacturing line with minimum disturbance, will always destroy the mode structure except for mono mode fibres. Such very thin mono mode fibres (approximately 100 μm diameter) will, however, have such high power loading for normal processing requirements that nonlinear optical losses will occur in them such as Brillouin , Raman and Rayleigh scattering (10). Some processes (e.g. spot welding thin sections or surface treatment) can stand the change in mode caused by passing down a fibre and so there is considerable effort being put into developing fibres for infra red 10.6 μm radiation as well as the existing silica fibres for YAG radiation.

Reduction of power loss:

A typical five axis robotic moving optics system will have around 7 optical components. If each had a loss due to absorption and surface scattering of around 2% this would total 13.8% loss. On a 5kW system this is equal to 660W loss, equivalent to a small laser. Power losses must be minimised because of this multiplication factor.

The reflectivity can be greatly improved if thin films of varying refractive indices are deposited on mirror surfaces (11).

Robust optics:

A Gaussian beam of typically 30mm diameter and say 10kW power will have a central power density of around 14W/mm2 which could generate around 5 C temperature difference between the centre and the edge of an edge cooled optic depending, of course, on the actual surface absorption of the component. This could be enough to crack ZnSe or other multicrystalline material if applied suddenly.

The avoidance of thermal shock has to be built into high powered beam guidance trains. This can either be done by keeping the optics warm with a pilot beam or avoiding the use of shock sensitive optics. Thus aerodynamic windows are the normal way of exciting the beam from a 10kW laser cavity; and metal optics are becoming the standard way of beam guidance and focussing. The development of parabolic mirrors capable of focussing and turning a beam through 45 is required. Cheap optics having such complex shapes will probably feature in the future. Looking further ahead there is the possibility of using high density gases to focus a laser beam. If the gases were flowing as in an aerodynamic window arrangement then a renewable nondestructible optic would be obtained. Such a device, if possible, would be very useful.

Pointing accuracy:

The focussed beam is around 100 µm and therefore the pointing accuracy must be better than this. A figure of around 25 µm is a suitable one on which to meditate. There are currently three basic guidance systems: gantry, articulated arm or fibre optic. Only the fibre optic can be mechanically placed the others rely on good software and no mechanical backlash. A new avenue through this problem may be found with fast software and feed back control of position from optical interference detectors working on the Michelson interferometer principle. In-process position sensors are under development.

Vibration:

It is some time since manufacturers first appreciated that the lightness of optical power does not mean that the guidance system can be dainty. A beam guidance system has to be rigid and rugged. Modern machinery with negligeable vibration is available; some improvements to master the problems with acceleration around corners are still required. Also vibration free movement at high speeds is difficult to attain. Such movement is going to be required when 10-20kW lasers are more common.

Swift guidance control:

The process control systems will be considered next, however there is one aspect of swift control which affects the design of the beam guidance train. It is that the optics should have minimal inertia. This means water cooling is not favored nor are heavy solid metal optics. What is left, from what is currently available, is very high reflectivity aircooled coated silicon wafer optics; similar systems will no doubt be developed in the future.

Process control systems:

Optical power can be directed and focussed easily it is also one of the chemically cleanest forms of energy available to industry. This together with the fact that the power from a laser can be controlled swiftly by electronic software signals means that the laser is the ideal partner for robotic or automatic manufacturing applications requiring energy.

Apart from this strong case for the growing use of lasers in automatic machinery there is the added advantage that optical processing has little electronic, magnetic, acoustic or thermal noise associated with it and thus allows, uniquely, in-process control from in-process sensors.

Numerous sensors are currently being developed, among these are the novel acoustic sensors developed at Imperial College (12). These sensors have shown that for certain lasers there is a ringing of the mirrors, which can indicate the state of a process while it is being performed.

There is a response time from the detectors which have and which will be developed. There must therefore be a fast response from the computer reading the signal and controlling the automatic machinery. There must also be a fast response by the laser, traversing table or beam focussing system to the control signal from the controlling computer. These three time steps must be fast for successful robotic control of the relatively fast processing which is possible with the laser. Thus the future will almost certainly see the development of extremely fast control systems possibly being ultimately dependent on glass computers driven by small semi-conductor laser signals. The end product will be fully automatic laser processing systems with 'intelligent' feed back.

Turning now to the possible future developments in the areas of process and application developments we will consider in turn the types of application, engineering improvements and material improvements:

Types of application:

The story of the current applications of lasers is a catalogue of the interactions between the enthusiasms of imaginative industrialists and those of inventive laser engineers. First someone perceives a problem and then someone engineers it.

It is difficult to list the current development areas which will feature in the future applications of lasers without trepassing on confidential areas. This is bound to be so because large sums of money are involved in the industrialisation of laser processes. However to list generally some areas of high activity would include the following:

Process area	General Nature of Development
Electronics	etching with u/v

depositing by LCVD
annealing
microwelding
brazing
drilling PCB

Xerography   high speed printers

Printing   high speed photogravure plate making

Cutting   High speed one off parts may lead to a
more
                            personalised market

Welding   High speed welding of thin material,
                            e.g. cans, cars etc. remarkable
                            quality is being attained in high
                            speed laser welding.

Surface treatments   localised hardening of many more parts
                            particularly in any moving machinery
                            from watches to machine tools the use
                            of shaped heat to harden parts by
                            single shots is a likely development.

Surface cladding   Many more parts will be clad instead
of using
                            expensive alloys or inserts or simply
                            allowing the piece to wear away.
                            Laser cladding has so little
                            disturbance of the part shape it is
                            quick and could be relatively cheap if
                            done in sufficient quantity.

Surface alloying   Similar to cladding.
                            There should be a dramatic change in
                            the type of materials used for many
                            applications. It is now possible
                            with laser surfacing methods to have a
                            cheap bulk material for strength and
                            almost any sophisticated surface alloy
                            suggested by metallurgical
                            considerations for optimal surface
                            properties. Between laser cladding
                            and alloying industry faces a
                            materials revolution but this is only
                            just being appreciated.

Surface glazing   Simply melting the surface of a
material
                            can produces a very fine
                            microstructure or a glassy metal.
                            This may find applications for
                            corrosion protection in environments

274

where alloying or cladding is unacceptable e.g. atomic power has limitiations on neutron captivity requirements.

Surface roughening surfaces

There is a possiblitiy of roughening by surface melting and blowing. Such surfaces could be useful for sugar rolls, metal stairways, sheet metal for cars (13) and the like.

Annealing applications

This was a big subject for laser a short time ago. Infact the laser pioneered the concept of rapid annealing of electronic circuits. However the laser having proved the success of the process was soon replaced by a very cheap hot wire!

This has been a feature of laser processing. Using the laser developes a process and this leads to the engineering of an alternative route. Transformation hardening has a number of examples of this.

Vapour Deposition lasers are

Finely focussed short wavelength capable of making fine prints by enhanced chemical deposition or localised pyrolysis of thermally sensitive chemicals. The application of these techniques to electronic circuits, machine tools or wear surfaces is currently being explored.

The 'cold working' which is sometimes possible with shorter wavelengths which operate by photolysis as opposed to pyrolysis allows clean print edges.

Medicine whole

The laser is clearly set to alter the field of surgery in all its aspects - external, endoscopic, brain and even cell surgery -. It is also set to alter the areas of cancer treatment by optically stimulated drugs as well as coronary surgery by laser heated cathetors. Dentistry and opthalmology will also be profoundly affected by this clinically clean, very precise tool.

Process Developments and understanding:

This area includes a better understanding of the nature of the interaction between laser light and a substrate. For example the understanding of the coupling of polarised light led to a doubling of the cutting speed and consequent improvement in the cut quality from a laser.

There will be further developments in this area but also in the handling of ancillary components such as the powder feed system developed by Imperial College and Quantum Laser Corp.(14)

An understanding of the correlation between mode structure and processing capability would almost certainly be fruitful.

Methods of using the plasma to the advantage of the process, such as the Welding Institutes plasma destruct jet (15) are areas ripe for development.

Control of surface reflectivity by optical feed back components (14) or surface treatments is another area in which developmemts can be expected.

Materials understanding:

The surfacing of different materials with the laser has led to many notable improvements in corrosion, wear and fatigue properties. The reason for these improvements is not always clear. Consider wear as an expample. It is often assumed that hardness is the main requirement for reduced wear. This is not always so. Some lathe beds are most resistant to wear when coated in nylon. It has been assumed that large carbide particles are required in hard facing alloys for good wear resistance. The laser produces very fine rapid quench structures having fine carbides and better wear properties. There is much metallurgy to learn from this new processing tool. Alloys once thought impossible for hard facing such as Fe/Cr/Mn/C can be made into a clad layer due to the rapid quench from laser processing, which avoids segregation problems (16). Such alloys it is to be noted do not contain strategic or expensive materials. Strengthening by gaseous surface alloying such as the nitriding of Ti in a nitrogen atmosphere has led to the possibility of forming TiN in other metals which contain Ti as an alloy ingredient (17).

Conclusion:

This is an incomplete analysis. Anything of this nature must be for the story of the world is full of surprises. However, one thing concerning the future that is certain is that we have much to learn and that many people will be involved and entertained in unravelling the threads that lie before us and weaving them into a pattern to suit our current

276

and future needs and in that fabric the laser has a part to play.

References:

1. Sona.A."High power lasers and their industrial application" Proc Laser 85 Opto Electronik conf. Munich 1985 ed Waidelich publ by Springer Verlag.

2. Shah.S., Steen.W.M. " Laser Oscillator/amplifier sytems" proc SPIE conf. Quebec June 1986 paper 668-06 to be published.

3. Belforte.D.A. "Robotic manipulation for laser processing" Proc SPIE conf Innsbruck Austria paper 650-38 April 1986 to be published.

4. Shock.W.,Witter.W.,Giesen.A.,Hall.T.,Hugel.H. "RF excitation of high power CO2 lasers" proc LIM4 conf, Paris June 1986 pp271-279 publ by IFS(publ)Ltd Kempston, Bedford, UK.

5. Wollermann-Windgasse.W."RF excited high power CO2 lasers for Industrial material processing" proc LIM4 conf., Paris June 1986 pp293-304 publ by IFS(publ) Ltd., kempston, Bedford, UK.

6. Hoffmann.P."Discharge Behaviour of a Rf excited high power laser at different excitation frequemcies" Proc SPIE conf Innsbruck Austria paper 650-03 April 1986 to be published.

7. P.Hoffmann "The Start of a new generation of CO2 lasers for industry" proc LIM2 conf Birmingham UK March 1985 pp210-208

8. Bramson.M.A. "Infra red radiation: A handbook of applications" Plenum Press, New York 1968.

9. Sharp.M.C., Steen.W.M. " Investigating process parameters for laser transformation hardening" paper 31 proc 1st Int conf on Surface Engineering June 1985 publ by the W.I. UK.

10. Weber.H.P." Non linear transmission of pulses" Proc SPIE conf Innsbruck Austria paper 650-15 April 1986 to be published.

11. Guenther.A.H., McIver.J.K."Considerations for the choice of optics for high power laser applications" Proc SPIE conf Innsbruck Austria paper 650-17 April 1986 to be published.

12. Weerasinghe.V.M., Steen.W.M. " Monitoring of laser material processing" Proc SPIE conf Innsbruck Austria paper 650-22 April 1986 to be published.

13. Crahay.J.,Renauld.Y.,Monfort.G., Bragard.A., "Present state of Development of the 'Lasertex' process" proc LIM4 conf., Paris June 1986 pp245-260 publ by IFS(publ) Ltd., kempston, Bedford, UK.

14. Weerasinghe.V.M., Steen.W.M., "Laser cladding with Pneumatic Powder Delivery", This volume

15. Pauley.J.T., Russel.J.D. US patent No 4 127 761 Nov 1978.

16. Singh.J.,Mazumder.J."Evolution of microstructures for laser clad Fe/Cr/Mn/C alloys" Proc SPIE conf Innsbruck Austria paper 650-32 April 1986 to be published.

17. Walker.A., Folkes.J., Steen.W.M., West.D.R.F. "Laser Surface Alloying of Titanium Substrates with Carbon and Nitrogen" Surface engineering Vol 1 No 1 pp23-30 1985

CHEMICAL PROCESSING WITH LASERS:

- Deposition
- Etching
- Oxide and Compound Formation

D. BÄUERLE   Lectures

Heading summarizes the title and contents of lectures delivered by Prof. D. Bäuerle, University of Linz, Austria.

Lecture notes were unexpectedly not available at the time of the course. Directors and Editors of Proceedings were given assurance by Prof. Bäuerle that a review paper concerning the lectures would be produced within months.

Delaying of proceedings publication could no longer be extended. Transcription from video-tapes recorded during the course did not guarantee desired quality and the lack of original illustrations. Editors and Publisher with regret were led to refrain from perfection and forced to drop the original determination of inclusion of the lectures material.

However, relevance of the topic suggested that a list of the papers referred during the course should be presented. Video-tapes could be made available for those interested on the topic.

References

G. Leyendecker   Laser Induced Chemical Vapor Deposition of Carbon
D. Bäuerle   Appl. Phys. Lett 39 (1981), 921-923
P. Geittner
H. Lydtin

D. Bäuerle   Ar Laser Induced Chemical Vapor Deposition of Si from Si_4H_4
P. Irsigler   Appl. Phys. Lett. 40 (1982), 819-821
G. Leyendecker
H. Noll
D. Wagner

G. Leyendecker   Rapid Determination of Apparent Activation Energies in
H. Noll   Chemical Vapor Deposition
D. Bäuerle   Journal Electromechanical Society 130(1983), 157-160
P. Geittner
H. Lydtin

W. Kräuter   Laser Induced Chemical Vapor Deposition of Ni by
D. Bäuerle   Decomposition of $Ni(CO)_4$
F. Fimberger   Appl. Phys. A 31(1983), 13-18

Soares, O.D.D., Perez-Amor, M. (eds), Applied Laser Tooling. ISBN-13: 978-94-010-8096-5
© *1987. Martinus Nijhoff Publishers, Dordrecht.*

280

D. Bäuerle   Production of Microstrutures by Laser Pyrolysis
       Mat. Res. Soc. Symp. Proc Vol 17 (1983), 19-28

D. Bäuerle   Laser Grown Single Crystals of Silicon
G. Leyendecker Appl. Phys. A 30 (1983), 147-149
D. Wagner

K. Piglmayer Temperature Distribution in CW Laser Induced Pyrolytic
J. Deppelbauer Deposition
D. Bäuerle   Mat. Res. Soc. Symp. Proc. Vol 29 (1983), 1-8, Boston

F. Petzold   Lateral Growth Rates in Laser CVD of Microstrutures
K. Piglmayer Appl. Phys. A 35 (1984), 155-159
W. Krauter
D. Bäuerle

D. Bäuerle   Laser - Induced Chemical Vapor Deposition
       Springer Verlag Series in Chemical Physics 39 (1984),
       166-182

D. Bäuerle   Materialbearbeitung mit Laserlicht Material Processing by
       Laser Radiation
       Laser und Optoelektronik 1 (1985), 29-36

D. Bäuerle   Chemical Processing with Lasers
       MRS - E, Strasbourg (1985)

Conclusion and Perspectives

The Machine-Tool industry is experiencing a great technological change led by American and Japanese hybridization of high technologies (machine software, control units, robotics, optics).

Laser Tooling was used and designed to induce a response to what answers Laser technologies can offer to the production systems. As machine-tools suffer from wearing, slow rate of production and the need for a large stock of tools it is likely that, in accelerated rate, power Lasers will become increasingly more valuable to Machine-Tool industry.

Laser design responds already with full compatibility to the sometimes harsh environment of the production sites(rugged enclosures, safety systems) but low unit cost, is as important as useful technical performance, and will have to be progressively met.

Robots and Lasers are being introduced as part of automation in cost-effective systems to improve productivity and to form part of the flexible production cells.

Chosen carefully Laser systems can offer unique advantages over other methods of manufacture with refined technology for a far from saturated market.

Laser systems and supporting hardware and software will continue as a frantic area of activity in parallel with the expected Laser applications growth in the future.

O.D.D. Soares

Soares, O.D.D., Perez-Amor, M. (eds), Applied Laser Tooling. ISBN-13: 978-94-010-8096-5
© *1987. Martinus Nijhoff Publishers, Dordrecht.*

Subject Index.

Soares, O.D.D., Perez-Amor, M. (eds), Applied Laser Tooling. ISBN-13: 978-94-010-8096-5
© 1987. Martinus Nijhoff Publishers, Dordrecht.